Urban Sustainability

The Urban Sustainability Book Series is a valuable resource for sustainability and urban-related education and research. It offers an inter-disciplinary platform covering all four areas of practice, policy, education, research, and their nexus. The publications in this series are related to critical areas of sustainability, urban studies, planning, and urban geography.

This book series aims to put together cutting-edge research findings linked to the overarching field of urban sustainability. The scope and nature of the topic are broad and interdisciplinary and bring together various associated disciplines from sustainable development, environmental sciences, urbanism, etc. With many advanced research findings in the field, there is a need to put together various discussions and contributions on specific sustainability fields, covering a good range of topics on sustainable development, sustainable urbanism, and urban sustainability. Despite the broad range of issues, we note the importance of practical and policy-oriented directions, extending the literature and directions and pathways towards achieving urban sustainability.

The series will appeal to urbanists, geographers, planners, engineers, architects, governmental authorities, policymakers, researchers of all levels, and to all of those interested in a wide-ranging overview of urban sustainability and its associated fields. The series includes monographs and edited volumes, covering a range of topics under the urban sustainability topic, which can also be used for teaching materials.

Abdulaziz I. Almulhim · Patrick Brandful Cobbinah

Urbanization Rethought

On Planning and Resilience in Saudi Arabia

Springer

Abdulaziz I. Almulhim
Department of Urban and Regional Planning,
College of Architecture and Planning
Imam Abdulrahman Bin Faisal University
Dammam, Saudi Arabia

Patrick Brandful Cobbinah
Faculty of Architecture, Building and Planning
The University of Melbourne
Melbourne, VIC, Australia

ISSN 2731-6483 ISSN 2731-6491 (electronic)
Urban Sustainability
ISBN 978-981-92-0117-4 ISBN 978-981-92-0118-1 (eBook)
https://doi.org/10.1007/978-981-92-0118-1

This Springer imprint is published by the registered company Springer Nature Singapore Pte Ltd.
The registered company address is: 152 Beach Road, #21-01/04 Gateway East, Singapore 189721, Singapore

To those shaping the future of Global South cities
and to all who plan, govern, and reimagine urban life under environmental limits.

Foreword

Urbanisation Rethought: On Planning and Resilience in Saudi Arabia advances our knowledge in meaningful and timely ways, particularly through its grounded engagement with the distinctive environmental, institutional, and developmental dynamics shaping contemporary urban transformation in Saudi Arabia. By situating questions of planning and resilience within this context, and within the intersecting realities of rapid growth, environmental constraint, institutional evolution, green infrastructure deployment, and megaproject-led development, the book foregrounds planning as a central mechanism through which resilience is shaped over time. It treats aridity, water scarcity, and heat not as background conditions but as defining parameters of urban governance. Importantly, it does not approach Saudi Arabia as an anomaly. Instead, it positions the country as a critical case through which broader assumptions about sustainable urbanization can be thoughtfully reconsidered, thereby deepening and diversifying the global urbanization discourse, challenging geographically narrow assumptions, and enriching theoretical and practical understanding alike. Its insights resonate well beyond national boundaries, offering lessons of broad relevance while remaining attentive to local specificity.

Urbanization has long been framed as a story of progress. Cities are celebrated as engines of economic growth, innovation, and cultural exchange, places where human ingenuity transforms complexity into opportunity. Yet in the early decades of the twenty-first century, this narrative is increasingly strained. Climate change, deepening inequality, ecological degradation, and the expanding power of digital technologies expose the limits of inherited models of urban development. The challenge before us is no longer whether urbanization should be rethought, but how profoundly and ethically that rethinking must occur as demonstrated in this book.

Urbanization is not merely a spatial or demographic process. It is a moral project. Cities materialize collective decisions about how people live together, how resources are distributed, how risks are shared, and whose futures matter. When these decisions are driven primarily by short-term gain, spectacle, or control, urbanization can intensify harm rather than well-being. Climate change has made this reality impossible to ignore. Extreme heat, flooding, water insecurity, and biodiversity loss are no longer abstract environmental concerns. They are everyday urban experiences that fall unevenly across populations, most severely affecting those with the least capacity to adapt. They reveal the consequences of planning systems that have privileged expansion over stewardship and efficiency over care.

Rethinking urbanization in the age of climate change therefore demands more than technical adjustment. It requires a reassessment of values. For much of the last century, urban development was guided by assumptions of abundance and mastery. Energy appeared cheap, land seemed limitless, and environmental impacts were treated as externalities to be addressed later. These assumptions have collapsed. Climate change confronts cities with thresholds that cannot be engineered away and with risks that compound across time and space. It calls for humility in the face of ecological limits and for planning approaches that acknowledge restraint as a form of responsibility rather than failure.

Alongside environmental stress, many societies are experiencing a quieter but equally destabilizing crisis of moral erosion. Trust in institutions has weakened, civic solidarity has frayed, and public debate has become increasingly polarized. Urban spaces reflect these tensions through exclusion, fragmentation, and the normalization of inequality. In such conditions, planning risks becoming either an instrument of narrow interests or a procedural exercise stripped of ethical purpose. To rethink urbanization is therefore also to confront questions of moral direction: what obligations cities owe to future generations, how urban systems balance individual freedom with collective well-being, and how public decision-making can be guided by care rather than convenience.

Rethinking urbanization in this context requires resisting simplistic narratives. Climate adaptation is neither purely technical nor purely political; it demands reconsidering the relationship between development ambition and ecological realism. Planning systems that prioritize speed, scale, and expansion without adequate regard for long-term environmental viability risk embedding structural vulnerabilities. Heat stress, water insecurity, and resource dependency are not peripheral concerns; they shape the durability of urban futures. Planning therefore assumes renewed significance. It mediates between development aspiration and environmental constraint, technological possibility and institutional capacity, and short-term gain and long-term resilience. In climate-stressed environments, decisions concerning land use, infrastructure, density, green infrastructure, and megaproject development accumulate over time, generating path dependencies that shape long-term adaptive capacity.

This reimagining begins with recognizing cities as socio-ecological systems rather than growth machines. Urbanization has often been measured in terms of density, productivity, and connectivity, while overlooking the flows of water, energy, materials, and governance capacity that sustain urban life. In arid regions, these flows are particularly fragile and resource-intensive, making long-term system viability a central planning concern rather than a peripheral one. Climate change exposes the vulnerability of these interdependent systems. Rethinking urbanization therefore means planning with ecological processes rather than against them, designing urban form and infrastructure that respond to climatic realities rather than defy them. In this light, resilience is not merely the capacity to recover from shocks, but the capacity to adapt and recalibrate over time through cumulative planning coherence across scales.

This is why context matters. Urban theory has often been shaped by temperate and comparatively resource-abundant environments where ecological constraints were less immediate. Yet cities shaped by aridity and extreme heat confront environmental limits more persistently and visibly. Rather than treating such contexts as peripheral, they should be understood as analytically generative sites for rethinking the relationship between urbanization, sustainability, and resilience in a warming world.

Climate stress, social anxiety, and economic volatility can encourage governance approaches that prioritize order and efficiency over inclusion and deliberation. Yet cities governed primarily through control are unlikely to be resilient in any meaningful sense. Resilience depends on trust, adaptability, and shared ownership of urban futures. Artificial intelligence can support these qualities if embedded within participatory planning processes and guided by clear ethical frameworks. It can erode them if it replaces public deliberation with automated decision-making.

The moral challenges facing urbanization are not simply matters of individual behavior. They are embedded in institutions, incentives, and dominant narratives of success. Urban development has frequently been framed as competition, encouraging excess, duplication, and symbolic projects that prioritize visibility over value. Climate change exposes the fragility of this logic. Rethinking urbanization requires a shift from competition to care, from accumulation to sufficiency, and from control to stewardship. Importantly, rethinking urbanization does not imply rejecting ambition or innovation. Cities remain sites of creativity and possibility. The task is to align ambition with responsibility and innovation with humility. Climate change demands bold action, but also restraint.

The future of urbanization is inseparable from the future of urban planning. Planning must move beyond procedural compliance and project facilitation toward a more reflective and normative role. It must ask not only what can be built, but what should be built. Not only how efficiently systems operate, but whom they serve and at what cost. In an era defined by climate change and emerging technologies, planning has a responsibility to act as a moral compass, guiding urban societies through uncertainty without abandoning democratic principles.

The task ahead is formidable, yet it is also an opportunity. Cities can become places where climate adaptation is pursued with justice, where technology is governed with care, and where moral responsibility is embedded in everyday decisions. Achieving this will require courage, critical reflection, and a willingness to challenge deeply embedded assumptions about growth, progress, and control.

This foreword invites readers to engage with urbanization not as a technical inevitability, but as a collective choice. The future of cities will be shaped by how societies respond to ecological limits, how they govern powerful technologies, and how they renew the moral foundations of urban life. Rethinking urbanization is therefore not simply an academic exercise. It is a necessary step toward cities that are resilient not only environmentally, but also ethically and humanly.

Professor Tan Yigitcanlar
QUT Urban AI Hub
Queensland University of Technology
Brisbane, Australia

Preface

Urbanization in Saudi Arabia represents one of the most concentrated and accelerated episodes of urban transformation in contemporary history. In recent decades, Saudi Arabia has moved from a landscape characterized by dispersed settlements and limited infrastructure to one defined by expansive metropolitan regions, complex infrastructural systems, and ambitious national development strategies. This transformation has unfolded not gradually but at extraordinary speed, reshaping space, institutions, and environmental relations in profound ways.

Yet Saudi urbanization cannot be fully understood through speed and scale alone. It unfolds within a structurally arid environment where extreme heat, chronic water scarcity, high evapotranspiration, and energy-intensive urban systems constitute enduring conditions rather than temporary constraints. Within this context, environmental limits are not peripheral considerations to be mitigated through technological adjustment; they are foundational parameters that shape urbanization patterns and influence. The primary aim of this book emerges from this intersection: what does urbanization mean when it occurs under conditions of persistent environmental constraint?

Much of the dominant literature on urban sustainability, resilience, and urban planning has been developed in temperate and resource-abundant contexts. The planning models and conceptual frameworks that circulate globally often assume ecological buffers that are comparatively generous. Concepts such as compact urbanism, green infrastructure, and climate adaptation strategies are frequently presented as universally transferable, with limited interrogation of how they perform under radically different climatic and governance regimes. When these models are applied to desert cities, however, their assumptions are unsettled. The Saudi context exposes the fragility of universal planning narratives and invites a more context-sensitive understanding of urban planning and development.

Similarly, despite the growing relevance of Saudi Arabia to debates on arid urbanization and climate-constrained development, there remains a notable scarcity of contemporary, integrative, book-length scholarship that systematically examines the intersection of planning, environmental constraint, and resilience in the Saudi context. Much of the available literature is either fragmented across journal articles or rooted in earlier development phases that predate the current era of rapid transformation and climate awareness.

This book was conceived, in part, as a response to this gap. It seeks to consolidate dispersed insights, critically synthesize existing knowledge, and provide a coherent analytical reference for scholars, students, practitioners, and policy institutions engaged with urban futures in Saudi Arabia and comparable arid regions. The book considers Saudi Arabia as an analytically generative context in which the tensions that underpin contemporary urbanization are intensified and rendered visible. Rapid urban growth collides with ecological constraint. Ambitious development visions confront institutional capacity. Megaprojects promise integration while amplifying scale related risk. Sustainability discourse expands even as resource pressures deepen. These contradictions are not unique to Saudi Arabia, but they are particularly complex, making the Saudi experience a revealing lens through which broader debates on urbanization and resilience can be rethought.

This book ultimately argues for a shift in how urban futures are imagined. Urban resilience cannot be reduced to technological upgrades, policy slogans, or iconic projects. It requires a fundamental reorientation of planning practice, one that takes environmental constraint seriously, embraces institutional learning, and acknowledges the cumulative consequences of spatial decisions. In arid contexts, margins for error are narrow. Planning, therefore, must become more deliberate, more integrated, and more attentive to long-term performance. At the heart of the book lies a clear argument: resilience in rapidly urbanizing arid environments is not solely a product of technological innovation or symbolic ambition. It is the cumulative outcome of planning decisions embedded within environmental realities and institutional structures. Planning is treated here not as a narrow regulatory function, but as a mediating practice through which development ambition and ecological limits are negotiated over time. Decisions about density, infrastructure alignment, governance coordination, and development sequencing accumulate, shaping patterns of vulnerability and adaptability that may endure for generations.

This book is written for scholars, planners, policymakers, and students seeking to understand how urban futures are shaped under conditions of environmental constraint. While rooted in the Saudi experience, the implications of this work extend well beyond national boundaries. As climate change intensifies heat stress and water scarcity across wider geographies, conditions once considered specific to desert regions are becoming increasingly relevant elsewhere. Arid cities are no longer peripheral to urban theory; they are at its frontier. The planning challenges confronting Saudi Arabia today are similar to many other cities, and this book contributes to a rethinking of urban resilience not as a slogan or policy aspiration, but as a long-term planning condition grounded in context, institution, and environment.

Dammam, Saudi Arabia | Abdulaziz I. Almulhim
Melbourne, Australia | Patrick Brandful Cobbinah

Acknowledgments

This book is the product of sustained intellectual engagement with the evolving urban realities of Saudi Arabia. It reflects years of research, dialogue, critical reflection, and ongoing commitment to understanding how planning, environmental constraint, and resilience intersect within arid urban contexts. The ideas developed in this volume have matured gradually through academic debate, empirical investigation, and careful theoretical reconsideration. We would like to acknowledge the broader scholarly community whose research, critiques, and intellectual contributions have shaped the contours of contemporary discussions on urbanization, sustainability, and resilience. Although this book advances its own analytical position, it is inevitably situated within a wider body of scholarship that has informed and challenged our thinking. We are especially appreciative of the professional exchanges, peer review processes, and academic environments that have provided space for the refinement of the arguments presented here. Scholarly work does not emerge in isolation; it develops through engagement, disagreement, and collective inquiry. In this sense, the book is part of an ongoing conversation about the future of cities under conditions of environmental constraint. We further acknowledge the support received from, and express our appreciation to the Urban and Regional Planning Department at Imam Abdulrahman Bin Faisal University, and the Faculty of Architecture Building and Planning at the University of Melbourne where Abdulaziz and Patrick work respectively. Finally, we acknowledge the patience and support of our families during the writing of this manuscript. The sustained effort required to produce a work of this scope is rarely visible, yet it depends upon understanding and encouragement beyond the academic sphere. Any remaining limitations or interpretations are solely the responsibility of the authors.

Dammam, Saudi Arabia — Abdulaziz I. Almulhim
Melbourne, Australia — Patrick Brandful Cobbinah

Competing Interests

The authors have no competing interests to declare that are relevant to the content of this manuscript.

About This Book

Urbanization Rethought: On Planning and Resilience in Saudi Arabia offers a critical examination of rapid urban transformation in Saudi Arabia through the interconnected lenses of planning, environmental constraint, and urban resilience. It situates Saudi urbanization within conditions of accelerated growth, structural aridity, and institutional transformation, arguing that these forces fundamentally reshape how cities must be planned, governed, and evaluated.

Rather than treating sustainability and resilience as universal policy templates, the book interrogates how these concepts operate when filtered through the realities of extreme heat, chronic water scarcity, energy-intensive infrastructure systems, and centralized governance structures. It advances a planning-led interpretation of resilience, emphasizing that long-term urban outcomes are shaped not by ambition or investment alone, but by cumulative spatial decisions embedded within ecological limits and institutional capacities.

The book integrates analyses of sustainable urbanization, the urbanization–environment nexus, climate policy and resilience framing, green infrastructure in arid environments, megaproject-led development, and metropolitan-scale implementation. Through empirical engagement with multiple Saudi cities and a detailed case study of the Dammam Metropolitan Area, it provides both conceptual depth and grounded insight into how urban futures are negotiated under conditions of environmental constraint.

A distinctive feature of this volume is its Saudi-first analytical perspective. The book uses the Saudi experience to question dominant paradigms derived from temperate and resource-abundant contexts. It foregrounds aridity not as a marginal condition but as a structural planning reality, one that increasingly anticipates the challenges facing cities in an increasingly warming world.

Urbanization Rethought is designed to serve multiple audiences. For researchers, it offers a coherent and integrative reference at the intersection of urban planning, resilience, and environmental constraint in arid regions. For graduate students in urban planning, geography, and environmental studies, it provides a structured theoretical and empirical framework suitable for advanced seminars on sustainability, climate adaptation, and Global South urbanism. For advanced undergraduate courses, selected chapters offer accessible yet rigorous material for examining rapid urbanization and planning in climate-constrained contexts. For policymakers and practitioners, the book presents analytically grounded reflections on how planning instruments, governance arrangements, and development strategies shape long-term resilience outcomes.

By positioning Saudi Arabia as an analytically generative context rather than an exceptional case, the book contributes to broader debates on how urban theory and planning practice must adapt to a century defined by environmental limits. It ultimately argues that rethinking urbanization in arid environments requires moving beyond symbolic ambition toward integrated, context-sensitive, and institutionally grounded planning approaches.

Prologue

Urbanization at the Edge of Limit

Urbanization has long been narrated as a story of expansion. Cities grow, economies diversify, skylines rise, and infrastructure networks extend outward. In much of urban theory, growth is assumed to be the default condition and planning the instrument that organizes it. Yet this narrative becomes unstable when urbanization unfolds at the edge of environmental limits.

What does it mean to urbanize where water is scarce, heat is extreme, and ecological margins are narrow?

What does resilience signify when stress is not episodic but constant?

What happens when ambition accelerates faster than institutional adaptation?

In arid environments, urbanization is not merely a process of spatial growth. It is a negotiation with constraint. Heat is not an anomaly; it is structure. Scarcity is not disruption; it is baseline. Infrastructure is not simply provision; it is survival. Under such conditions, planning cannot remain procedural or reactive. It becomes the medium through which environmental realities and developmental aspirations are reconciled.

Saudi Arabia represents one of the most concentrated arenas where these tensions converge. Within a single generation, cities have expanded at extraordinary speed, supported by unprecedented investment and national transformation agendas. Yet this expansion unfolds within a desert ecology that resists abstraction. Environmental constraint does not retreat in the face of ambition. It accumulates consequences.

The significance of this context lies not in its exceptionality but in its clarity. Here, the contradictions of contemporary urbanization are intensified. Growth collides with ecological limits. Megaprojects promise integration while amplifying systemic risk. Sustainability is invoked even as resource demand escalates. Resilience becomes policy language while vulnerability deepens in less visible ways.

These tensions are not uniquely Saudi. They are increasingly global. As climate change expands zones of heat stress and water insecurity, cities once considered climatically stable confront new realities. What appears today as a desert condition may tomorrow become a widespread urban condition. In this sense, arid urbanization is not peripheral to urban theory; it is anticipatory.

To rethink urbanization under these conditions requires more than technological optimism or rhetorical commitment to sustainability. It requires confronting the cumulative consequences of planning decisions. Density, land use, infrastructure alignment, governance coordination, and development sequencing are not neutral instruments. They are long-term commitments embedded in environmental systems that do not easily forgive error.

Urbanization Rethought emerges from this recognition. It approaches planning not as a regulatory afterthought, but as a mediating practice between ambition and limit. It treats resilience not as recovery from shock, but as the capacity to operate within persistent environmental stress. It situates Saudi Arabia not as a peripheral case study, but as a revealing site through which foundational assumptions about urban development can be questioned.

This book invites the reader to move beyond inherited narratives of urban expansion and toward a more deliberate inquiry:

How should cities grow when ecological margins are narrow?
How should planning function when uncertainty is structural?
How can ambition be recalibrated without extinguishing possibility?

Urbanization at the edge of limit is not a story of retreat. It is a story of recalibration. It demands intellectual humility, institutional learning, and planning practices that acknowledge that not all futures are equally viable.

If urban theory is to remain relevant in a warming century, it must take aridity seriously. It must recognize that environmental constraint is not an obstacle to be engineered away, but a condition to be integrated into the very logic of development.

The chapters that follow explore this terrain. They examine how sustainability, resilience, green infrastructure, megaprojects, and governance operate when filtered through the realities of desert urbanization. They argue that the future of cities under climate stress will be determined less by symbolic ambition and more by the cumulative wisdom of planning decisions made today.

Urbanization at the edge of limit is not merely a Saudi question. It is an emerging global condition.

This book is an invitation to rethink it.

Dammam, Saudi Arabia — Abdulaziz I. Almulhim
Melbourne, Australia — Patrick Brandful Cobbinah

Contents

Part I Conceptual Foundations

1 Urbanization in Arid Contexts: Planning, Constraint, and Resilience 3
1.1 Why Urbanization Rethink Matters? 3
1.2 Urbanization and Planning Under Arid Conditions: The Saudi Challenge 5
1.3 Why Aridity Changes Everything: Planning in Desert Environments 7
1.4 Saudi Arabia as an Analytical Urban Laboratory 8
1.5 Toward a Consolidative Inquiry into Saudi Urbanism, Planning, and Resilience 9
1.5.1 Rethinking Planning Through the Saudi Experience 9
1.5.2 Sustainability and Rapid Urbanization: Limits of Growth-Oriented Models 10
1.5.3 Framing Urban Resilience: From Policy Discourse to Planning Practice 10
1.5.4 Green Infrastructure and Adaptation: Context-Sensitive Planning Responses 11
1.5.5 Planning Scale and Ambition: Megaprojects as a Resilience Test 12
1.5.6 Governance, Institutions, and the Reality of Implementation 12
1.6 Contribution of the Book 13
1.7 Structure of the Book 15
1.8 Concluding Framing: Rethinking Urban Futures in Arid Contexts 15
References 16

2 Saudi Arabian Cities in Sustainable Urbanization Dialogue 19
2.1 Introduction 19
2.2 Sustainable Development and Urbanizing Global South Cities 21
2.3 Research Methods 24
2.3.1 Study Area Context 24
2.3.2 Data Collection and Analysis 25
2.4 Urbanization and Sustainable Development in Saudi Arabia 25
2.4.1 The Urbanization–Urban Planning Collision: Why Sustainable Development Matters in Saudi Arabian Cities 25
2.4.2 Situational Barriers to Sustainable Urbanization in Saudi Arabia 28
2.4.3 Enabling Sustainable Urbanization in Saudi Arabia 35
2.5 Discussion and Conclusion: The Agency of Sustainable Urbanization in Saudi Arabia 39
References 41

3 Sustainable Development Fantasies and Realities in Saudi Arabian Cities: Urbanization-Environment Analysis 49
3.1 Introduction 49
3.2 Environment-Urbanization Nexus: Toward Sustainability 52
3.3 Research Approach 55
3.3.1 Research Context 55
3.3.2 Data Collection and Analysis 55
3.4 Urbanization-Induced Environmental Quality Threats in Saudi Arabia's Cities 56
3.4.1 Understanding Pollution (Water, Air, and Land) in Urbanizing Country 56
3.4.2 Urbanization and Water Resource Depletion 57
3.4.3 Urbanization's Implications on Urban Heat Islands 58
3.4.4 Sea-Level Rise and Urban Flooding 58
3.4.5 Low Urban Agricultural Productivity 59
3.4.6 Urban Sprawl and Slum Development 60
3.4.7 Deforestation and Wetland Destruction 61
3.4.8 Urbanization as a Threat to Public Health 62
3.5 Discussion: Realities of Sustainable Development in Saudi Urbanization 63
3.6 Conclusion 66
References 67

Part II Planning and Resilience in Practice

4 Mediating Urbanization and Sustainable Development in Saudi Arabia: Urban Planning Pivotality and Stewardship 75
4.1 Introduction 75
4.2 Urbanization-Sustainable Development Linkages: A Focus on Saudi Arabia 77
4.3 Research Approach 80
4.3.1 Study Area Context 80
4.3.2 Data Collection and Analysis 82
4.4 Urban Planning and Sustainable Urbanization Narrative in Saudi Arabia 85
4.4.1 Overview of the Saudi Urban Planning Context 86
4.4.2 Central Sustainability Inadequacies in Saudi Arabia's Urbanization Management 88
4.4.3 Sustainable Urbanization: The Challenge of Inclusivity in Saudi Arabia 90
4.4.4 The Threat of Environmental Problems to Sustainable Urbanization in Saudi Arabia 91
4.5 Discussion: Toward Sustainable Urbanization: The Centrality of Planning 93
4.6 Conclusion 95
References 97

5 Desiring Resilience in Saudi Arabian Cities: On Climate Change and Urban Policy 103
5.1 Introduction 103
5.2 Why Should Cities in the Global South Desire Urban Resilience? 105
5.3 Analytical Context and Approach 109
5.3.1 Analytical Context 109
5.3.2 Data Gathering and Analysis Approach 110
5.4 Urban Resilience in Saudi Arabia: Analysis of Climate Change and Urban Policy Readiness 113
5.4.1 Saudi Arabian Cities' Readiness for Future Uncertainty 113
5.4.2 The Resilience of Saudi Arabian Cities 116
5.5 Key Contributions 127
5.6 Discussion and Conclusion: Desiring Resilience in Rapidly Urbanizing Saudi Cities 128
References 129

6 **Green Infrastructure: What It Means for Climate Adaptation and Planning in Saudi Arabia** ... 137
6.1 Introduction: Green Infrastructure and Urban Resilience in Arid Contexts ... 137
6.2 Rethinking Green Infrastructure Beyond Urban Greening ... 140
6.3 Climate Adaptation Imperatives in Arid Urban Environments ... 142
6.4 Planning Green Infrastructure Under Water Scarcity ... 144
6.5 Saudi Cities as Stress Tests for Green Infrastructure Strategies ... 147
6.6 Green Infrastructure and Urban Resilience: Planning Implications ... 150
6.7 Conclusion: Toward Context-Sensitive Green Infrastructure Planning ... 153
References ... 154

Part III Governance, Transformation, and Urban Futures

7 **Managing Megaprojects and Building Resilience in Saudi Arabian Cities** ... 159
7.1 Introduction ... 159
7.2 Megaprojects in Planning Theory and Urban Resilience Discourse ... 162
7.3 Saudi Arabia as a Megaproject-Driven Urban Context ... 165
7.4 Megaprojects and Urban Resilience: Tensions and Trade-Offs ... 168
7.5 Planning Megaprojects Under Arid and Climate-Constrained Conditions ... 170
7.6 Governance, Planning Capacity, and Implementation Challenges ... 174
7.7 Rethinking Megaprojects as Planning Instruments for Resilience ... 178
7.8 Conclusion: From Iconic Development to Resilient Urban Futures ... 181
References ... 183

8 **Advancing the Pursuit of Urban Resilience: Profiling Dammam Metropolitan Area** ... 185
8.1 Introduction ... 185
8.2 The Urban Nexus Framework as an Urban Resilience Approach ... 188
8.3 Methods ... 190
8.3.1 DMA: Geography and Demography ... 190
8.3.2 Dataset Collection Sources for the Urban Nexus Assessment ... 192
8.3.3 LULC Mapping ... 193
8.3.4 Measurement of Land Surface Temperature ... 193

8.4 Results: Urban Resilience Analysis in DMA 195
8.4.1 Land Use Land Cover (LULC) Change Dynamics 195
8.4.2 Urbanization Pressures on Natural Ecosystems in DMA 197
8.4.3 Spatial LST Distribution During 1993–2023 199
8.4.4 Influence of Natural Ecosystem Loss on LST in DMA 203
8.4.5 Analysis of Air Pollutant Concentration in DMA 203
8.5 Contours of Urban Resilience in DMA: Urban Nexus Positionality 205
8.6 Discussion 209
8.6.1 Scholarly Insights 209
8.6.2 Policy Prescriptions 211
8.7 Conclusion 212
References 214

9 Planning for Sustainable Urban Futures: Rethinking Urbanization and Resilience in Saudi Arabia 223
9.1 Urbanization under Arid Constraint: Core Insights 223
9.2 Planning as a Mediating Practice, Not a Technical Tool 224
9.3 Sustainability, Growth, and the Limits of Reactive Urbanism 225
9.4 Rethinking Urban Resilience in Climate-Constrained Cities 225
9.5 Green Infrastructure and Megaprojects as Conditional Planning Instruments 226
9.6 Governance, Institutions, and the Implementation Gap 227
9.7 What the Saudi Experience Reveals for Arid Urban Futures 228
9.8 Final Reflections 229

Index 231

About the Authors

Abdulaziz I. Almulhim is an internationally recognized scholar in urban sustainability and climate change. He is Associate Professor of Urban Sustainability at Imam Abdulrahman Bin Faisal University, Saudi Arabia. His research critically examines urban planning under conditions of rapid urbanization, climate stress, and environmental constraint. He has published over 70 peer-reviewed articles and is recognized among the world's top 2% of scientists (Stanford University). He serves as a Lead Author for the IPCC Special Report on Climate Change and Cities (2027) and sits on the editorial boards of leading international journals. He has led and edited major scholarly volumes published by Springer, Palgrave Macmillan, and Elsevier. His research focuses on climate change adaptation, sustainable smart cities, and the transformation of rapidly urbanizing regions toward climate-positive futures. His work integrates theory, policy, and practice to advance resilient and sustainable urban transitions globally.

Patrick Brandful Cobbinah is an Associate Professor of Urban Planning at the University of Melbourne and a leading academic commentator on urban planning in Global South with over 130 academic publications. Recognized among the world's top 2% of scientists by Stanford University for three consecutive times (2023, 2024, 2025), and as a national rising star of research in 2021 by the Australian Research Magazine, his research focuses on urban planning and climate change in Global South and the outcomes of inappropriate planning practices and theories on Global South cities. His work continues to unsettle and challenge the existing geopolitics of knowledge production in planning and climate management by providing a deeper understanding and enriching Global South-focused theorization on planning and climate change. He is the Senior editor of *Journal of Urban Affairs* and the editor of four flagship books: *Handbook on Planning and Climate Change Adaptation* (Edward Elgars), *Reimagining Urban Planning in Africa* (Cambridge University Press), *Sustainable Urban Futures in Africa* (Routledge), and *The Geography of Climate Change Adaptation in Urban Africa* (Palgrave-Macmillan). He is a lead member of the OECD's Working Group on "Accelerating Africa's built environment towards a sustainable and resilient future."

Part I
Conceptual Foundations

Chapter 1
Urbanization in Arid Contexts: Planning, Constraint, and Resilience

1.1 Why Urbanization Rethink Matters?

Among urban studies and planning scholars, consensus is materializing that sustainable urbanization is a sensible strategy to both manage increasing urban population and generate socio-economic benefits as well improved livelihoods across cities (Almulhim & Cobbinah, 2023; Ochoa et al., 2018; Tan et al., 2016). From debates on planetary urbanization (Brenner, 2018), peripheral urbanization (Caldeira, 2017) to sustainable urbanization, cities are deemed sustainable when there is adequacy of resources to support their management and governance as well as providing alternative ways to address and recover from eventualities, stresses and shocks via the offerings of long-term opportunities to maintain and improve the urban functionality (Nijkamp & Perrels, 2018; Sodiq et al., 2019). Yet urbanization across many Global South cities has been defined by unregulated informality, poverty, housing crisis, infrastructure inadequacy, and significant governance failures.

Despite profound criticism of the urbanization narratives in the Global South (see Sridhar & Mavrotas, 2021), the concept nevertheless remains prominent and underscores the discourse on transformative urban development that creates environmental awareness and sustainable livelihoods. This discourse extends across new, overlapping, variants of urban studies and development logics such as resilience, climate change and sustainable development. In addressing a variety of social, economic, and environmental issues associated with urbanization, rethinking how the concept captures novel or alternative ways in which urban development occurs, particularly in the Global South, that differs from conventional urbanization narrative is timely and tenable.

The ongoing rapid urbanization and urban development debate in the Global South revolves around several dimensions, which are key to understanding sustainable forms and challenges of *doing* urbanization. There is growing awareness that the intended long-term effects of climate change depend on how urbanization is managed (Bulkeley, 2013; Stone et al., 2012). Urbanization is conceived in this book

A. I. Almulhim and P. B. Cobbinah, *Urbanization Rethought*, Urban Sustainability,
https://doi.org/10.1007/978-981-92-0118-1_1

as the process of urban transformation characterized by changing morphology of the cityscape and increasing concentration of people (Cobbinah et al., 2015). Complicating matters further, climate change, which is a significant global threat, is considerably shaping the urbanization challenges worldwide, particularly in the Global South, as it is often described as an existential threat and urban development predicament in vulnerable regions of the world where the capacity to provide sustained economic benefits and livelihood enhancement to urbanizing population is limited while maintenance of ecosystem integrity has become a distant hope (Cobbinah & Addaney, 2022). To address the increasing danger and impacts of climate change on cities, urbanization needs to be rethought and repositioned in the Global South to deliver benefits that protect the natural environment, support poverty eradication, and prioritizes urban residents in urban development.

However, there is mounting evidence that many cities across the Global South, particularly in Asia and Africa, often fail to recognize the enormity of climate change in urban development and policy (Yuen, & Kumssa, 2010). In many countries, climate change relevance in urbanization dialogue in terms of policy and action is, in terms of Cobbinah and Finn (2023), mere rhetoric and this contributes to increased climate change impacts across the cities. There are many examples, particularly in the Middle East, Africa, South East Asia, and Latin America where the impacts of climate change such as heatwaves, droughts, and flood events have intensified and become widespread due to limited or lack of attention to understanding the complexity of urbanization and climate change in the process of urban development (Finn & Cobbinah, 2023; Reyer et al., 2017; Waha et al., 2017), and where informal urban expansion is rarely regulated or considered by city management (Finn & Cobbinah, 2023). Commenting on the climate change complexity in rapidly urbanizing cities, urbanization understanding needs to evolve and transcend population growth and spatial expansion and focus on the complex interactions within and without such as decision-making models, infrastructure capabilities, and the knowledge of critical urban issues (e.g., climate change) among residents.

As claims of urban vulnerability due to climate change have come under critical scrutiny, urban studies and planning scholars have advocated for an effective urban planning regime to address the urbanization complications in managing climate change impacts (Blakely, 2022; Cobbinah et al., 2019). However, they have raised issues with presenting urban planning as passive or reactional model which does not have influence shaping, driving, and advancing sustainable urbanization. In many Global South cities, urban planning remains elitist, exclusionary, and ineffective denying urban residents the benefits of sustainable urbanization—i.e., inclusive, participatory, future-oriented, proactive, and adaptable (Almulhim & Cobbinah, 2024a). More importantly, sustainable urbanization requires urban planning initiatives which help build management and leadership skills and strengthen local institutions thereby driving the local social and environmental economy with strong ties to collaborative networks, including formal social ties which may facilitate access to resources and the actual ability address critical urban challenges (Cobbinah et al., 2019). When urban planning becomes strengthened, resilience initiatives also become collaborative, genuine, and impactful in terms of preserving natural and cultural heritage and delivering improved socio-economic outcomes (Almulhim & Cobbinah, 2024a).

The consideration of climate change, urban planning, and resilience in the urbanization discourse, however, is not an undemanding task, and available studies on them often scratch the surface becoming limited in scope and analysis. This lack of comprehensive understanding of urbanization in contemporary city context fails to question the scale and scope of the concept and ignores that urban development is complex and contextual, implying varying mobilities, interdependencies, and collaborations that transcend simplistic application or analysis of planning and resilience issues within rapidly urbanizing and changing city climate. Hence, this book focuses on rethinking urbanization within the complex caricature of urban development, using Saudi Arabia as a case of deeply embedded urban governance struggles, striking climate change challenges, complex urban planning terrain, and inspiring stories of resilience. We argue that without discussing urbanization within the evolving complex settings, sustainable urban development across cities in the Global South is likely to be incomplete, divided, and unattainable.

Overall, this book distinguishes three dimensions of the sustainable urbanization that are central to urban development discourse: (1) climate change as a complicating factor in urbanization debates, (2) recognition of urban planning as an important tool and process in managing urbanization, and (3) urban resilience strongly relies on urbanization management. The ways in which the experience of Saudi Arabian cities resonates with these dimensions are explored in this book.

1.2 Urbanization and Planning Under Arid Conditions: The Saudi Challenge

Urbanization is often theorized as a process involving population growth, spatial expansion, and morphological transformation (Almulhim & Cobbinah, 2023). In arid environments, however, urbanization unfolds not as an open-ended trajectory, but as a condition structured by environmental constraint. Heat, water scarcity, and climatic uncertainty are not external variables to be managed, but foundational parameters that shape planning rationalities, institutional arrangements, and urban futures.

Urbanization in Saudi Arabia has unfolded at a pace and scale that place it among the most intensive urban transformations of the late twentieth and early twenty-first centuries. Within a few decades, the Saudi Arabia has transitioned from predominantly rural and small urban settlements to a highly urbanized society characterized by large metropolitan regions, extensive infrastructure networks, and rapidly expanding urban footprints (UN-Habitat, 2019). This transformation has not been gradual or evolutionary; rather, it has been driven by accelerated economic development, demographic change, and state-led investment, producing cities that have grown faster than the planning frameworks traditionally used to manage them.

Yet urbanization in Saudi Arabia cannot be understood solely through the lens of speed or scale. It unfolds within a structurally arid environment, where extreme heat, chronic water scarcity, and high energy dependence are not episodic challenges but enduring conditions that shape everyday urban life (UN-Habitat, 2019). These environmental constraints fundamentally condition how cities function, how infrastructure performs, and how urban risks accumulate over time. As such, urbanization in Saudi Arabia poses a distinctive planning challenge: how to manage rapid urban growth under conditions where ecological limits are tightly drawn and margins for error are narrow.

This book argues that the significance of Saudi urbanization lies not in its exceptionality, but in its analytical clarity. The Saudi context exposes, in intensified form, many of the contradictions that underpin contemporary urban development elsewhere: the tension between growth and sustainability, between ambition and environmental constraint, and between policy vision and planning capacity. What distinguishes Saudi Arabia is that these tensions are not gradually unfolding; they are concentrated and accelerated, making them visible in ways that challenge conventional planning assumptions.

Although comprehensive planning frameworks were introduced from the 1970s onward, rapid population growth, centralized governance, and oil-driven development priorities frequently constrained planning capacity, resulting in fragmented and sprawling urban expansion (Almulhim & Cobbinah, 2024a). The consequences of this reactive approach are increasingly evident in patterns of urban sprawl, infrastructure stress, socio-spatial fragmentation, and environmental degradation. These outcomes have prompted renewed attention to planning as a central instrument for managing urban transformation rather than as a technical exercise confined to land-use regulation.

In recent years, sustainability, resilience, and adaptation have become central pillars within Saudi Arabia's national policy frameworks, particularly in the context of climate action and sustainable development (Government of Saudi Arabia, 2025). Cities are now framed as sites where climate risks, extreme heat, flooding, water scarcity, and dust storms must be anticipated and managed. However, the incorporation of these concepts into planning practice remains uneven. This book contends that the challenge facing Saudi cities is not a lack of policy ambition, but a deeper need to rethink how planning mediates between rapid urbanization and environmental constraint.

The core premise of this book is that urban resilience in Saudi Arabia cannot be achieved through the uncritical adoption of global models or best practices. Instead, resilience must be understood as both a planning process and outcome that emerges from context-sensitive decisions about urban form, infrastructure systems, governance arrangements, and institutional capacity. By examining Saudi urbanization through this lens, the book seeks to move beyond descriptive accounts of growth and toward a critical analysis of how planning can, and must, respond to the realities of arid urban environments.

1.3 Why Aridity Changes Everything: Planning in Desert Environments

Aridity is not merely a climatic descriptor; it represents a structural environmental condition that shapes the possibilities and limits of urbanization. In desert and climate-vulnerable environments, water scarcity, extreme heat, and climate-related stresses impose constraints that affect urban functionality, infrastructure systems, and long-term sustainability outcomes (Almulhim & Cobbinah, 2025). Unlike temperate regions, where environmental limits may be encountered episodically, arid cities operate under continuous environmental stress. Planning in such contexts must therefore contend with a fundamentally different baseline.

In Saudi Arabia, aridity intersects with rapid urban growth to produce a set of planning challenges that are both acute and cumulative. Urban expansion increases demand for water, energy, and cooling, while rising temperatures intensify heat stress and reduce outdoor liveability. Infrastructure systems designed to support urban growth become increasingly resource-intensive, reinforcing dependence on energy-heavy water production and cooling technologies (Government of Saudi Arabia, 2025). These dynamics create feedback loops in which urbanization amplifies environmental stress, and environmental stress, in turn, raises the costs and risks of urban development.

Climate change acts as an intensifier of these existing conditions rather than as a wholly new threat. Projections of rising temperatures and increasing climate variability heighten concerns about long-term urban resilience (Government of Saudi Arabia, 2025). In this context, planning for resilience is not about preparing for rare shocks but about managing persistent stress. This distinction is critical, as it challenges resilience frameworks that emphasize recovery from discrete events rather than adaptation to chronic environmental pressure.

Much of the contemporary urban sustainability discourse has been shaped by experiences and policy models emerging from Global North cities, with limited comparative attention to how sustainability and resilience strategies perform across diverse geographic and climatic contexts Almulhim & Cobbinah, 2025). Concepts such as green infrastructure, compact urbanism, and nature-based solutions are frequently framed as transferable planning approaches, although their effectiveness may vary significantly depending on local environmental and socio-ecological conditions. In arid environments, however, such approaches may produce unintended consequences if environmental constraints are not fully accounted for. For example, urban greening strategies that rely on intensive irrigation can exacerbate water scarcity, while compact urban forms may intensify heat exposure if not carefully designed.

The Saudi experience underscores the importance of contextualizing urban sustainability strategies within arid environments characterized by water scarcity, extreme heat, and high resource demands (Almulhim & Cobbinah, 2023). In this regard, the transferability of planning models across geographic contexts requires careful reassessment, as urban sustainability framings and responses vary significantly depending on local environmental and institutional conditions (Almulhim &

Cobbinah, 2025). Rather than assuming universal applicability, sustainability interventions must be evaluated in relation to long-term environmental performance and infrastructural resilience within specific climatic realities (Almulhim & Cobbinah, 2023).

This book adopts the position that arid environments should not be treated as marginal or exceptional within urban theory. On the contrary, as climate change expands the spatial reach of heat stress and water scarcity, the planning challenges faced by desert cities are likely to become increasingly relevant elsewhere. Saudi Arabia thus offers a vantage point from which to interrogate the adequacy of existing planning paradigms under conditions of environmental constraint.

1.4 Saudi Arabia as an Analytical Urban Laboratory

Framing Saudi Arabia as an analytical urban laboratory does not imply that its cities represent models to be replicated or exceptions to be explained away. Rather, it recognizes that the Saudi context brings into sharp relief the interactions between rapid urbanization, environmental constraint, and planning capacity. The combination of accelerated growth, centralized governance, and arid conditions creates a setting in which planning decisions have amplified consequences, both positive and negative.

Saudi cities are characterized by high levels of urban concentration, with major metropolitan areas such as Riyadh, Jeddah, and Dammam acting as focal points for population growth, economic activity, and infrastructure investment (UN-Habitat, 2019). This concentration intensifies urban pressures while also creating opportunities for strategic planning interventions. At the same time, the scale of development and the speed of implementation place significant demands on planning institutions, testing their ability to coordinate across sectors, manage complexity, and integrate long-term resilience considerations.

A defining feature of the Saudi urban context is the central role of the state in directing urban development. Planning frameworks are embedded within national development agendas, including the National Spatial Strategy, and successive five-year development plans, while major infrastructure investments and spatial priorities are largely coordinated through centralized institutional structures (UN-Habitat, 2019). From a governance perspective, such alignment can facilitate decisive action and resource mobilization, but it may also constrain flexibility and limit opportunities for incremental learning. The book approaches this institutional configuration analytically, examining how governance structures shape planning outcomes rather than treating them as fixed or normative.

Importantly, the Saudi experience complicates binary distinctions between "developed" and "developing" urban contexts. While Saudi cities exhibit advanced infrastructure systems and substantial state-led investment capacity (UN-Habitat, 2019), they also face environmental pressures, spatial imbalances, and governance challenges commonly associated with rapidly urbanizing contexts (Almulhim &

Cobbinah, 2023; UN-Habitat, 2019). This hybridity makes Saudi Arabia a particularly valuable case for rethinking urban theory and planning practice beyond conventional categories.

By situating Saudi Arabia at the center of the analysis, the book reverses the typical direction of knowledge transfer. Instead of applying global theories to a local case, it uses the Saudi experience to question the assumptions embedded in those theories. The chapters that follow examine how concepts such as sustainability, resilience, green infrastructure, and megaproject-led development operate when filtered through the realities of arid urbanization.

1.5 Toward a Consolidative Inquiry into Saudi Urbanism, Planning, and Resilience

1.5.1 Rethinking Planning Through the Saudi Experience

The trajectory of urbanization in Saudi Arabia highlights the need to reassess the role of urban planning. Evidence suggests that planning institutions have often struggled to effectively guide rapid urban expansion, with growth frequently outpacing regulatory and spatial frameworks (Almulhim & Cobbinah, 2023). Urban development frequently followed economic imperatives, sectoral investments, and political priorities, with planning frameworks adapting reactively to decisions made elsewhere. This historical marginalization of planning has left a visible imprint on Saudi cities, reflected in fragmented urban form, infrastructure stress, and limited coordination across land use, transport, and environmental systems.

In recent years, planning in Saudi Arabia has been repositioned as a strategic governance instrument aligned with national transformation agendas and sustainability objectives (UN-Habitat, 2019). This shift reflects growing recognition that uncoordinated growth is no longer viable under conditions of environmental constraint and climate stress. Yet this repositioning remains incomplete. Planning institutions are now expected to manage complex urban challenges, ranging from climate adaptation to megaproject integration, often without commensurate expansion in authority, capacity, or institutional coherence. The Saudi experience thus highlights a critical tension: planning is increasingly invoked as a solution, but not always empowered to act as one.

This book argues that planning should be understood not as a static regulatory framework, but as a dynamic institutional practice that shapes urban vulnerability and resilience over time. In arid contexts, planning decisions related to density, land use, infrastructure alignment, and governance arrangements have long-term implications for resource consumption and environmental exposure. The Saudi case underscores how the absence of proactive, integrated planning can lock cities into development pathways that are difficult and costly to alter. Conversely, it also illustrates the potential of planning-led approaches to reorient urbanization toward more resilient trajectories when institutional conditions allow.

1.5.2 Sustainability and Rapid Urbanization: Limits of Growth-Oriented Models

Sustainability is a central concept in contemporary urban discourse, yet its meaning and application remain contested in rapidly urbanizing contexts. In Saudi Arabia, sustainability has often been framed in terms of economic diversification, infrastructure provision, and service delivery, with environmental considerations incorporated unevenly. Rapid urban growth has generated significant pressures on land, water, and energy systems, raising questions about whether high-speed urbanization can be reconciled with ecological sustainability and long-term social and infrastructural resilience (Almulhim & Cobbinah, 2023; UN-Habitat, 2019).

Chapters 2 and 3 of this book engage directly with this question by examining the environmental and socio-spatial consequences of accelerated urbanization in Saudi cities. Patterns of urban sprawl, infrastructure overload, and resource depletion reveal the limits of growth-oriented development models that prioritize expansion over integration. These patterns are not unique to Saudi Arabia, but they are intensified by arid conditions that magnify the environmental costs of inefficient urban form. In this sense, Saudi cities expose the structural fragility of sustainability approaches that fail to engage with context-specific constraints.

The book advances the argument that sustainability in Saudi Arabia cannot be achieved through reactive policy adjustments or isolated interventions. Instead, it requires a proactive planning orientation that anticipates environmental limits and integrates sustainability considerations into core planning instruments. Compact urban form, transit-oriented development, and inclusive planning mechanisms are not presented as generic solutions, but as planning strategies whose relevance and performance must be evaluated within the specific ecological and institutional context of Saudi cities. By situating sustainability within a planning-led framework, the book challenges narratives that treat sustainability as an outcome of economic growth rather than as a condition shaped by spatial and institutional choices.

1.5.3 Framing Urban Resilience: From Policy Discourse to Planning Practice

As climate risks have become more visible, resilience has emerged as a prominent concept in Saudi urban policy discourse (Almulhim & Cobbinah, 2024b). Cities are increasingly framed as sites of vulnerability to extreme heat, flooding, water scarcity, and dust storms, prompting the adoption of resilience-oriented strategies at national and municipal levels. While this policy attention marks an important shift, it also raises critical questions about how resilience is conceptualized and operationalized within planning practice.

This book adopts a critical stance toward resilience as a policy keyword, arguing that resilience must be understood as a planning outcome rather than a rhetorical objective. Chapter 5 examines how resilience is framed in Saudi urban policy and identifies gaps between conceptual ambition and implementation capacity. These gaps are particularly evident in the translation of climate assessments into land-use decisions, zoning regulations, and infrastructure investments. Without such translation, resilience risks remaining an abstract aspiration rather than a tangible urban condition.

In arid environments, resilience is less about recovery from discrete shocks and more about managing chronic stress. Extreme heat, water scarcity, and environmental degradation exert continuous pressure on urban systems, shaping everyday risks and long-term vulnerability. Planning responses that prioritize short-term robustness without addressing underlying exposure may inadvertently increase vulnerability over time. The Saudi experience thus illustrates the need to move beyond risk management toward adaptive urban transformation, where planning frameworks are designed to evolve in response to environmental change.

1.5.4 Green Infrastructure and Adaptation: Context-Sensitive Planning Responses

Green infrastructure has gained prominence as a planning response to climate change and urban resilience, often promoted as a multifunctional solution capable of addressing heat stress, flooding, and environmental degradation simultaneously. However, its applicability in arid and water-scarce environments remains contested. Chapter 6 critically examines the role of green infrastructure in Saudi cities, questioning assumptions about its universal transferability.

The book argues that green infrastructure cannot be understood independently of the water–energy–urban planning nexus that defines arid urban systems. Interventions that appear environmentally beneficial in temperate contexts may impose significant resource burdens in desert cities, particularly when long-term maintenance and irrigation requirements are considered. This does not imply that green infrastructure is irrelevant in Saudi Arabia, but rather that it must be selective, strategic, and planning-led.

By situating green infrastructure within the realities of extreme heat, water scarcity, and governance constraints, the book reframes it as one component within a broader adaptation toolkit rather than as a stand-alone solution. This perspective aligns with the book's broader argument that resilience emerges from integrated planning decisions rather than from isolated interventions. The Saudi case thus contributes to a more nuanced understanding of how adaptation strategies must be tailored to environmental context.

1.5.5 Planning Scale and Ambition: Megaprojects as a Resilience Test

Large-scale urban development has become a defining feature of contemporary urbanization in Saudi Arabia (Alsayel et al., 2025). Megaprojects, operating at unprecedented scales of space, investment, and governance, are reshaping urban regions and redefining planning practice. Chapter 7 examines these developments through the lens of urban resilience, highlighting the tensions and trade-offs inherent in megaproject-led urbanism.

The book conceptualizes megaprojects as planning instruments rather than as inherently positive or negative interventions. Their scale enables integrated infrastructure provision and strategic coordination, but it also amplifies risk, resource intensity, and institutional complexity. In arid and climate-constrained contexts, these dynamics are particularly consequential. Planning decisions embedded within megaprojects shape urban vulnerability for decades, making the alignment between ambition and environmental realism critical.

By analyzing megaprojects within the broader planning system, the book challenges narratives that equate scale with resilience. It argues that megaprojects can contribute to resilient urban futures only when they are embedded within robust planning frameworks that prioritize adaptability, integration with existing cities, and long-term operational capacity. The Saudi experience illustrates how megaprojects function as stress tests for planning institutions, revealing both their strengths and limitations.

1.5.6 Governance, Institutions, and the Reality of Implementation

Across the themes examined in this book, rapid urbanization, sustainability, resilience, green infrastructure, and megaproject-led development, governance, and institutional capacity emerge as a unifying concern. Planning outcomes in Saudi Arabia are shaped not only by spatial strategies or policy objectives, but by the institutional arrangements through which decisions are made, coordinated, and implemented. The effectiveness of planning interventions therefore depends as much on governance structures as on technical design.

Saudi Arabia's urban governance landscape is characterized by a complex interplay between national development strategies, sectoral ministries, and locally implemented planning mechanisms (UN-Habitat, 2019). While centralized authority can enable strategic alignment and resource mobilization, it can also generate coordination challenges and institutional rigidity. Planning institutions are frequently required to operate across administrative boundaries and policy domains, managing interactions between land use, infrastructure, environmental regulation, and economic development. These demands are intensified in the context of megaprojects and climate adaptation, where decisions carry long-term implications for urban resilience.

This book approaches governance analytically rather than normatively. It does not assume that particular institutional models are inherently superior, nor does it advocate wholesale decentralization or centralization as a panacea. Instead, it examines how governance arrangements shape the capacity of planning systems to respond to environmental constraint and urban complexity. Chapters 4, 5, 7, and 8 demonstrate that gaps between planning vision and operational reality often arise not from a lack of policy intent, but from misalignments between institutional responsibilities, regulatory frameworks, and implementation capacity.

A critical insight emerging from the Saudi experience is that operational resilience cannot be achieved through planning vision alone. Long-term maintenance, monitoring, and adaptation require institutional continuity and learning. In arid environments, where infrastructure performance and environmental management are ongoing challenges, the sustainability of planning interventions depends on governance systems that extend beyond project delivery toward long-term stewardship. This recognition reinforces the book's central argument: that planning for resilience is as much an institutional challenge as a spatial one.

1.6 Contribution of the Book

This book makes several interrelated contributions to the literature on urban planning, sustainability, and resilience, with a particular focus on arid and climate-constrained urbanization. First, it advances a Saudi-first analytical perspective that treats the Saudi Arabia not as a peripheral or illustrative case, but as a substantive urban context through which broader planning debates can be critically reconsidered. By foregrounding Saudi urbanization, the book challenges the widespread tendency to generalize from temperate, resource-abundant settings and highlights the central role of environmental constraint in shaping planning outcomes.

Despite the growing recognition of Saudi Arabia's relevance to debates on urbanization in arid environments, there remains a notable absence of contemporary, book-length scholarship that systematically examines Saudi urbanization through an integrated planning and resilience lens. Existing English-language books on Saudi cities are limited in number and scope. They are largely dated, produced several decades ago, and predominantly structured as edited volumes reflecting the concerns of earlier development phases, such as institutional formation, descriptive urban profiles, and foundational planning challenges. These works predate the emergence of climate change, urban resilience, green infrastructure, and megaproject-led urbanism as defining forces shaping contemporary cities.

More recent scholarship on Saudi urbanization has appeared primarily in the form of journal articles, policy reports, and sector-specific studies. While valuable, this body of work remains fragmented, thematically narrow, and analytically dispersed. It does not provide a coherent, book-length framework capable of capturing the cumulative environmental, institutional, and planning challenges associated with rapid urbanization under arid and climate-constrained conditions. In Arabic, although several books address individual cities, most notably Riyadh, or focus on specific aspects of urban growth, these works similarly lack an integrated and contemporary planning perspective that engages systematically with sustainability, resilience, and environmental constraint at the national scale.

Against this backdrop, this book represents one of the first collectively authored, comprehensive examinations of urbanization in Saudi Arabia to bring together planning theory, urban resilience, environmental limits, and contemporary development instruments within a unified analytical framework. Rather than offering descriptive accounts or policy narratives, the book critically interrogates how Saudi cities have been planned, governed, and transformed, and how these processes shape long-term urban vulnerability and resilience. In doing so, it positions Saudi Arabia not as an exceptional or peripheral case, but as a revealing urban context through which broader questions about planning, resilience, and urbanization in arid environments can be rethought.

Second, the book contributes to planning theory by emphasizing planning as a mediating institutional practice between ambition and constraint. Instead of framing planning as a purely technical or regulatory activity, it conceptualizes planning as a process through which environmental limits, governance structures, and development priorities are negotiated over time. This perspective bridges discussions of sustainability, resilience, and governance, offering a more integrated understanding of how planning decisions accumulate and shape long-term urban outcomes.

Third, the book contributes to the literature on urban resilience by critically examining how resilience is operationalized in arid contexts. It challenges narratives that present resilience as a universally transferable framework, arguing instead for context-sensitive, planning-led approaches that explicitly recognize environmental limits and institutional realities. Through its analysis of green infrastructure and megaproject-led development, the book demonstrates how commonly promoted resilience strategies may perform differently under conditions of water scarcity, extreme heat, and resource dependence.

Finally, the book offers an empirically grounded contribution to arid urban studies. Drawing on multiple Saudi cities and a detailed metropolitan case study, it illustrates how planning challenges and resilience outcomes vary across spatial scales and institutional settings. This empirical depth strengthens the book's theoretical arguments and enhances its relevance to scholars, practitioners, and policymakers working in arid and rapidly urbanizing regions. In addition to its scholarly contribution, the book is structured to support teaching across undergraduate and graduate programs in urban planning, geography, and environmental studies, where there is a growing need for context-specific, theory-informed material on urbanization in arid environments.

1.7 Structure of the Book

The book is organized to guide the reader from foundational conceptual questions about urbanization, planning, and environmental constraint in Saudi Arabia to more focused analyses of sustainability, resilience, and planning instruments, before concluding with broader reflections on urban futures in arid contexts. This chapter establishes the conceptual and contextual foundations of the book. It situates Saudi urbanization within the conditions of aridity, rapid growth, and institutional transformation, and frames planning and urban resilience as central analytical lenses through which contemporary urban challenges are examined. Chapter 2 examines whether rapid urbanization can be sustainable in the Saudi context, focusing on patterns of urban growth, infrastructure stress, and socio-spatial inequality in major cities. Chapter 3 deepens this analysis by exploring the urbanization–environment conundrum, highlighting the environmental implications of accelerated urban growth and the limitations of development models that prioritize socio-economic objectives over ecological considerations.

Chapter 4 shifts attention to the role of urban planning, critiquing its historically limited influence and examining recent efforts to reposition planning as a central driver of urban transformation. Chapter 5 builds on this discussion by analyzing how urban resilience is framed within Saudi climate policy and urban planning, identifying gaps between conceptual ambition and implementation capacity. Chapter 6 focuses on green infrastructure and climate adaptation in arid urban environments, offering a critical assessment of their potential and limitations in Saudi cities. Chapter 7 examines megaproject-led urban development as a defining feature of contemporary Saudi urbanization, analyzing its implications for planning capacity, governance, and urban resilience. Chapter 8 concludes the empirical analysis with a metropolitan-scale case study of the Dammam Metropolitan Area, illustrating how resilience-oriented planning unfolds in practice across different urban zones and institutional settings. Chapter 9 brings the book to a close by synthesizing insights across the preceding chapters. It reflects on the implications of the Saudi experience for planning and urban resilience in arid environments and outlines directions for future research, policy, and planning practice.

1.8 Concluding Framing: Rethinking Urban Futures in Arid Contexts

This book argues that rethinking urbanization in Saudi Arabia requires moving beyond singular solutions and toward integrated, planning-led approaches that engage seriously with environmental constraint and institutional capacity. Rapid urbanization in arid environments presents challenges that cannot be resolved through policy ambition or technological innovation alone. Instead, it demands sustained attention to how planning decisions shape urban form, infrastructure systems, and governance arrangements over time.

By situating Saudi Arabia at the center of the analysis, the book demonstrates how arid cities function as critical sites for interrogating the adequacy of contemporary planning paradigms. The lessons emerging from the Saudi experience are not confined to the Saudi Arabia; they speak to broader debates about urban resilience in a warming world where heat stress and water scarcity are becoming increasingly widespread. In this sense, the book invites readers to rethink not only urbanization in Saudi Arabia, but the assumptions that underpin planning and resilience discourse more generally.

References

Almulhim, A. I., & Cobbinah, P. B. (2023). Can rapid urbanization be sustainable? The case of Saudi Arabian cities. *Habitat International, 139*(102), 884. https://doi.org/10.1016/j.habitatint.2023.102884

Almulhim, A. I., & Cobbinah, P. B. (2024a). Urbanization provocateur: Reaching urban planning-led development in Saudi Arabia. *Land Use Policy, 147*, Article 107365.

Almulhim, A. I., & Cobbinah, P. B. (2024b). Framing resilience in Saudi Arabian cities: On climate change and urban policy. *Sustainable Cities and Society, 101*, Article 105172. https://doi.org/10.1016/j.scs.2024.105172

Almulhim, A. I., & Cobbinah, P. B. (2025). On urban sustainability: Urbanization and climate change collision. *Sustainable Development*. https://doi.org/10.1002/sd.70382

Alsayel, A., Fransen, J., De Jong, M., & Almatar, K. M. (2025). Riyadh: King of gigaprojects can Riyadh's dramatic place rebranding realize path-breaking and profound urban transformations? *Cities, 163*, Article 106022. https://doi.org/10.1016/j.cities.2025.106022

Blakely, E. J. (2022). *Urban planning for climate change*. Lincoln Institute of Land Policy.

Brenner, N. (2018). Debating planetary urbanization: For an engaged pluralism. *Environment and Planning D: Society and Space, 36*(3), 570–590.

Bulkeley, H. (2013). *Cities and climate change*. Routledge.

Caldeira, T. P. (2017). Peripheral urbanization: Autoconstruction, transversal logics, and politics in cities of the global south. *Environment and Planning D: Society and Space, 35*(1), 3–20.

Cobbinah, P. B., & Addaney, M. (Eds.). (2022). *Sustainable urban futures in Africa*. Routledge.

Cobbinah, P. B., Asibey, M. O., Opoku-Gyamfi, M., & Peprah, C. (2019). Urban planning and climate change in Ghana. *Journal of Urban Management, 8*(2), 261–271.

Cobbinah, P. B., Erdiaw-Kwasie, M. O., & Amoateng, P. (2015). Africa's urbanisation: Implications for sustainable development. *Cities, 47*, 62–72.

Cobbinah, P. B., & Finn, B. M. (2023). Planning and climate change in African cities: Informal urbanization and 'Just' Urban transformations. *Journal of Planning Literature, 38*(3), 361–379.

Finn, B. M., & Cobbinah, P. B. (2023). African urbanisation at the confluence of informality and climate change. *Urban Studies, 60*(3), 405–424.

Government of Saudi Arabia. (2025). *Second Nationally Determined Contribution (NDC) report: 2025 Submission to UNFCCC*.

Nijkamp, P., & Perrels, A. (2018). *Sustainable cities in Europe*. Routledge.

Ochoa, J. J., Tan, Y., Qian, Q. K., Shen, L., & Moreno, E. L. (2018). Learning from best practices in sustainable urbanization. *Habitat International, 78*, 83–95.

Reyer, C. P., Adams, S., Albrecht, T., Baarsch, F., Boit, A., Canales Trujillo, N., Thonicke, K., et al. (2017). Climate change impacts in Latin America and the Caribbean and their implications for development. *Regional Environmental Change, 17*(6), 1601–1621.

Sridhar, K. S., & Mavrotas, G. (Eds.). (2021). *Urbanization in the Global South: Perspectives and challenges*. Routledge.

Sodiq, A., Baloch, A. A., Khan, S. A., Sezer, N., Mahmoud, S., Jama, M., & Abdelaal, A. (2019). Towards modern sustainable cities: Review of sustainability principles and trends. *Journal of Cleaner Production, 227*, 972–1001.

Stone, B., Vargo, J., & Habeeb, D. (2012). Managing climate change in cities: Will climate action plans work? *Landscape and Urban Planning, 107*(3), 263–271.

Tan, Y., Xu, H., & Zhang, X. (2016). Sustainable urbanization in China: A comprehensive literature review. *Cities, 55*, 82–93.

UN-Habitat. (2019). *Saudi cities report 2019. Future Saudi cities program.* https://unhabitat.org/sites/default/files/2020/05/saudi_city_report.english.pdf

Waha, K., Krummenauer, L., Adams, S., Aich, V., Baarsch, F., Coumou, D., Schleussner, C. F., et al. (2017). Climate change impacts in the Middle East and Northern Africa (MENA) region and their implications for vulnerable population groups. *Regional Environmental Change, 17*(6), 1623–1638.

Yuen, B., & Kumssa, A. (Eds.). (2010). *Climate change and sustainable urban development in Africa and Asia*. Springer Science & Business Media.

Chapter 2
Saudi Arabian Cities in Sustainable Urbanization Dialogue

2.1 Introduction

It has been widely reported that urbanization has emerged as a key urban policy concern in recent decades (Ahani & Dadashpoor, 2021; Almulhim & Cobbinah, 2022; Elmqvist et al., 2021; He & Zhang, 2022; Hölscher & Frantzeskaki, 2021) despite the persistence of its conceptual opacity (Cobbinah et al., 2015). Here, some (e.g., McGranahan & Satterthwaite, 2014) contend that the global importance of urbanization is frequently framed and lost in inconsistent country-level ascribed meanings. According to this viewpoint, urbanization has been alternatively and variously interpreted based on the positionality and level of development of nation states. As a result, in many ways, what is considered as urban in one context is not in another (McGranahan & Satterthwaite, 2014). While this is not necessarily a problem, it distorts global-level comparison and assessments. Many have also described urbanization as a phenomenon and a process focusing on both the act of increasing population concentration in cities driven by natural increase and in-migration, and the spatial morphological changes that characterize such population growth (Cobbinah et al., 2015; Elmqvist et al., 2021; Hölscher & Frantzeskaki, 2021). Within this context, urbanization is often described in dimensions such as rapid, where the rate of population concentration exceeds the planning and management capacity of cities (World Cities Report, 2022); and abnormal, where the process does not lead to socio-economic growth of city citizens, and where the expected timing and conditions of progress associated with urbanization do not occur, stifling the capacity of city managers to adequately plan for the change (see Obeng-Odoom, 2010). Whereas opponents contemplate urbanization as a problematic form of urban growth, especially in developing countries (see Camhis, 2006), its advocates consider it as an opportunity to spur the socio-economic growth of cities and improve the livelihoods of city citizens, having the capacity to generate agglomeration opportunities (see Lange, 2017). Considering these doubts and debates, it may not be irrational to say that urbanization is only an imaginary hassle and an urban growth stunt.

A. I. Almulhim and P. B. Cobbinah, *Urbanization Rethought*, Urban Sustainability,
https://doi.org/10.1007/978-981-92-0118-1_2

Yet, in reality, urbanization is producing severe city management pathologies, such as uncoordinated urban development, destruction of ecologically sensitive areas, and increased poverty and deprivation across many developing countries (see Andreola Serraglio et al., 2019; Pilehvar, 2021; Bai et al., 2017; Lu et al., 2022; Serraglio et al., 2020). The urbanization phenomenon in developing countries is defined by an agenda of urban planning chasing urban development, dysfunctional urban spaces, and unsustainable urban ecosystems (Bai et al., 2017; Pilehvar, 2021). Amidst this complicated discussion, Alshebli (2018) and Hölscher and Frantzeskaki (2021) assert that urbanization in developing countries is mostly inconsistent with sustainable city development. Obeng-Odoom (2010) analyzes various descriptions and contentions of urbanization in developing countries as poverty-driven, premature, and abnormal.

At present, urbanization in developing countries, especially those in Asia and Africa, has become a significant impediment for city planners and managers (Cobbinah et al., 2015; Hölscher & Frantzeskaki, 2021; Poku-Boansi, 2021). Urban planning distortions amidst unprecedented urbanization rates have collided in producing a disorder, epitomized by increased poverty, slum growth, traffic congestion, urban sprawl, etc., that requires urgent response by researchers, governments, and city managers to address and deliver sustainable urban development. Although efforts to address this phenomenon of rapid urbanization are often pursued in developing countries, particularly in Asia, they are mostly centered on non-Arab countries (e.g., Cohen, 2006; Kiamba, 2012; Potter & Lloyd-Evans, 2014; Stiftel, 2021), with limited research focusing on Arab countries (e.g., Abubakar & Dano, 2020; Almulhim & Cobbinah, 2022; Almulhim et al., 2022; Elimam, 2022). However, Arab countries remain among the fastest-growing and most urbanized regions in developing countries (General Authority for Statistics (GAS), 2022; UN-Habitat, 2019), with huge urbanization–sustainable development challenges (Abou-Korin & Al-Shihri, 2015). For example, Saudi Arabia is one of the most urbanized countries in the world, with its urbanization levels expected to surpass 90% by 2030 (UN-Habitat, 2019). Considering the dangers of, and comparatively inadequate understanding of, inter alia, sustainable urbanization, it is expected that the discrepancy between the advancement made toward sustainable urban development and the complications of urbanization will remain wide in many Arab cities, particularly those in Saudi Arabia (see Aljaddani et al., 2022), which has frequently provoked concern and increased activities of many global agencies including UN-Habitat (see UN-Habitat, 2019).

With a growing interest of global agencies in the urbanization and sustainable development discourse through international agendas such as the sustainable development goals (SDGs) and the New Urban Agenda, the urgency for an in-depth assessment of how sustainable urbanization can be realized in the Arab world is increasing. It has become important to analyze, in a variety of ways, the situational limitations of sustainable urbanization in Saudi Arabian cities and to compare them with the mushrooming literature on the issue globally. Until these deficiencies in knowledge on sustainable urbanization are addressed, current urban planning and management inadequacies may persist in Saudi and other Arabian cities. In response to this knowledge gap, this chapter poses the question as to whether sustainable urbanization can be achieved in Saudi Arabian cities. It addresses this question by (i) analyzing the relationship between sustainable development and urbanization in the Global South, (ii) examining the urbanization–urban planning collision in Saudi Arabian cities, and (iii) assessing the situational limitations to and enablers of sustainable urbanization in Saudi Arabia.

This chapter is organized as follows: First, we provide a critical literature review of sustainable development and urbanization. Specifically, we discuss the history and evolution of sustainable development philosophy and how it is shaping and framing urban development discourses, particularly in the Global South. Second, we briefly explain the research context and methods used in undertaking this analysis. We then present the results in three parts: (a) review the literature on the foundations of the urbanization–urban planning collision to explain the efficacy or otherwise of urban planning in managing urbanization. In particular, in this section we explain the opposing viewpoints on urbanization studies to understand the various ways in which urban growth is deeply intertwined with urban planning and sustainable development; (b) we examine the current urbanization literature in Saudi Arabia. We argue that while Saudi Arabia is rapidly urbanizing with considerable urban management challenges, framing of sustainable development is lacking; (c) we draw on previous literature to argue for contextualization of urbanization–urban planning within sustainable development thinking in Saudi Arabia. Finally, the discussion and conclusion section of the chapter is presented.

2.2 Sustainable Development and Urbanizing Global South Cities

The first recorded use of the concept "sustainable development" was at the 1972 UN Conference on the Human Environment. The subsequent publication of the 1987 Brundtland Report, Our Common Future, and the 1992 UN Conference on Environment and Development (UNCED) instigated and substantially accelerated debate and discussion on sustainable development. According to Du Pisani (2006), the emergence of the sustainable development concept was strongly linked to an increasing consciousness of an impending environmental predicament and remained a foundational force, especially in the twenty-first century. The outcome has been a proliferation of studies and articles in which sustainable development is discussed predominantly within a developed country setting (e.g., Carruthers, 2001; Harlow et al., 2013; Tyburski, 2008), although other studies in the Global South where the complications seem complex are rapidly emerging (e.g., Cobbinah et al., 2011; Fukuda-Parr & Muchhala, 2020; Fulekar et al., 2014).

In many urban studies and planning literature, the sustainable development discourse, as some writers (e.g., Liu et al., 2014; Rydin, 2012; Yigitcanlar & Teriman, 2015) explain, has focused on creating livable and functional urban spaces at a time when most cities in the Global South are confronted with unprecedented rapid urbanization, transforming urban morphologies in an uncertain climate. In this case, it is somewhat unsurprising that sustainable development has become a catchphrase for the urban development narrative in the Global South.

Regrettably, across many Global South countries, city managers are struggling to plan and manage rapid urbanization and its associated impacts, a situation that is threatening sustainable urban development (Camhis, 2006; Cobbinah & Finn, 2022; Cobbinah et al., 2015). Urbanization represents a significant societal change in the contemporary era (Bai et al., 2017; Pilehvar, 2021), mostly resulting from a natural increase in the population, reconfiguration of urban and regional systems, or rural-to-urban migration (Alshebli, 2018). At the regional, national, and international levels, urbanization has significant and complex impacts on ecology and economic disparity (Elmqvist et al., 2021; Hölscher & Frantzeskaki, 2021). Currently, the urban population constitutes more than half of the global population (4.3 billion people according to Christley et al. [2021]), and this is expected to grow to nearly two-thirds by 2030. Cities are set to expand at a rate of almost 95% in the Global South (World Cities Report, 2022). Urban expansion, excessive energy consumption, lack of essential infrastructure and services, traffic jams, and climate change are some of the impacts of rapid urbanization on sustainable development (Abubakar & Dano, 2020; Almulhim & Cobbinah, 2022; Elimam, 2022).

Urban planning, also referred to as spatial planning or land-use planning, is the core element for urban development, and for empowering regional and national authorities to manage the complexity of urban expansion, global warming, and migration (Adelina et al., 2020; Cilliers et al., 2021; Kramer et al., 2022). However, it remains weak, inadequate, and mostly unresponsive across multiple Global South countries (Alderton et al., 2021; Khakee, 2020). The conceptualization of urban planning as a policy-driven method of governance does not necessarily result in the creation of a new model for planning and management (Asadzadeh et al., 2022), but it does highlight the interconnectedness of the concepts of governance management and urban planning along with the necessity for more effective spatial governance and planning to address the inadequacies of the current systems in the Global South (Alderton et al., 2021; Khakee, 2020) and to promote sustainable development.

And yet even though the sustainable development agenda has become common and imperative (United Nations, 2015), only a relatively small proportion of the mushrooming literature has addressed the question of the urbanization–sustainable development relationship in the Global South (e.g., Arab countries), although it is often discussed in siloes (see Saghir & Santoro, 2018). Without an understanding of and research into the relationships and dynamics of urbanization–sustainable development interactions, it may be difficult, if not impossible, for cities in the Global South to realize sustainable development that both promotes urban functionality and advances healthy urbanism.

Undoubtedly, there is literature published on urbanization and urban planning in Global South countries. Jorge Hardoy, David Satterthwaite, and Diana Mitlin, for instance, have been studying urbanization and urban planning in developing countries of Africa and Asia for some time, and have been joined by others in recent times (e.g., Ahani & Dadashpoor, 2021; Cobbinah et al., 2015; Cohen, 2006; Gambe et al., 2023; He & Zhang, 2022; Kiamba, 2012; Ochoa et al., 2018; Potter & Lloyd-Evans,

2014). The international community has promoted further activities, most notably from the United Nations, whose agenda "Sustainable Development Goals, Sendai Framework for Disaster Risk Reduction and the New Urban Agenda" has activated a significant space for debates and actions on the future of global cities, particularly those in the Global South (Alqahtany & Aravindakshan, 2021; Sandoval & Sarmiento, 2020; Stiftel, 2021; UN-Habitat, 2016a; United Nations, 2018). Many other contributions to sustainable urban development in the Global South abound; yet, to a significant extent, the relationship between urbanization, urban planning and sustainable development has had a peripheral role in the debate on cities in the Global South, despite the considerable implications of this relationship on urban survival, healthy urbanism, and functional cities. In fact, it may be reasonable to argue, extrapolating from Cobbinah et al. (2015) and Ye and Niyogi (2022), that for the majority the most pressing concern of cities in the Global South is managing rapid urbanization and climate change, which is central to producing sustainable urban spaces. In the view of Cobbinah et al. (2015) and Deb and Okulicz-Kozaryn (2023), unplanned urban development and deprivation are critical consequences of rapid urbanization without sustainable development in the Global South, and that managing them would substantially enhance the sustainable development image of cities in the Global South. Despite this, much of the discourse on urban sustainability in the Global South, regardless of how well-intentioned or important, is happening without sufficient attention paid to the urbanization–sustainable development relationship (see Potter & Lloyd-Evans, 2014).

Despite this limitation, there is an evolving literature on the urbanization–sustainable development relationship in the Global South (e.g., Cohen, 2006; Kiamba, 2012; Rana & Parves, 2011). The challenge remains, however, that few writers have endeavored to analyze this relationship from an Arabian perspective. Meanwhile the Arabian context remains a powerhouse for rapid urbanization (see Fig. 2.1).

As a result, several questions on the urbanization–sustainable development relationship remain unresolved in many Global South countries across the Arabian world. For example, how do sustainable development agendas influence urbanization management in Arabian cities? And what have been the outcomes of urbanization on the urban planning response to sustainable development in Arabian cities? To help address these questions, this chapter uses Saudi Arabia to analyze the interplay of urbanization, urban planning, and sustainable development in framing the future of Arabian cities.

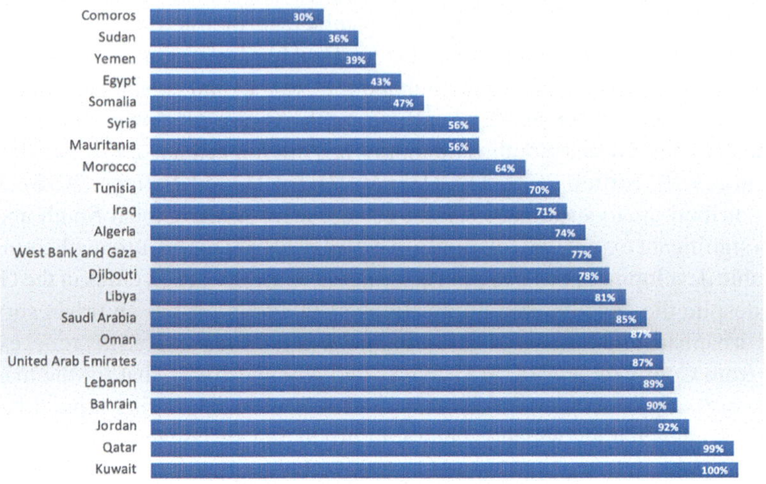

Fig. 2.1 Urbanization levels in the Arab World countries (Authors' computation using data extracted from World Bank, 2020)

2.3 Research Methods

2.3.1 Study Area Context

In 2021, Saudi Arabia had around 34 million residents, with 83.8% of them living in cities (GAS, 2022). Based on the country's predicted average annual growth rate of 2.7%, this figure is expected to increase to 37.6 million by 2025. The largest nation in the Middle East, and the 14th largest in the world, Saudi Arabia's total surface area is 2.15 million km2, occupying 80% of the Arabian Peninsula (CIA, 2019). Its climate is characterized by extremely scanty and erratic rain (average annual rainfall of 70.5 mm), arid winds, and extreme evapotranspiration, contributing to a lack of freshwater and diminished vegetation (Al-Wabel et al., 2020). The annual average temperature is rising by about 0.75 °C, which, by 2040, is expected to reach 2.2–2.7 °C, above the 1.5–2.0 °C global average (Addas & Maghrabi, 2021). Although the nation has 365,716 acres of coral reefs, 8649 acres of mangroves, and 2700 thousand hectares of medium forest, which accounts for more than half of its land area, which is made up of arid mountain ranges (Abubakar & Dano, 2020). In 2022, the country's population density was 16.67 persons per m2, and its median age was 32.4. The country is divided legally into 13 regions. Its gross domestic product (GDP) is expected to reach $700.1 billion in 2020, accounting for 0.62% of the global market, making it the 15th-richest country in the world, and the fourth-richest Arab nation (Saudi Arabia Population Statistics, 2022).

2.3.2 Data Collection and Analysis

This chapter is based on an extensive review of literature. Data collection was conducted by analyzing relevant primary literature and secondary data (Sileyew, 2019). First, relevant literature was collated by using keywords such as "sustainable cities," "urban planning," "urbanization," "community resilience," "sustainable urban development," "urban governance," and "climate adaptation." Google Scholar, Science Direct, and Crossref were used for the detailed searches. This chapter compiled online sources from journals, books, scientific documents, meetings, and websites. A thematic data analysis was used, and a two-step iterative method employed as the analysis technique. Although some materials fit into more than one category, the resources were first arranged according to the search criteria. In the absence of sufficient material, each resource was thoroughly searched by iterating through it to confirm the relevant themes or discover fresh perspectives. Ultimately, the themes were combined by determining how they related to one another and where they diverged or converged (Vourvachis & Woodward, 2015). Overall, about 100 publications were analyzed for this chapter.

2.4 Urbanization and Sustainable Development in Saudi Arabia

2.4.1 The Urbanization–Urban Planning Collision: Why Sustainable Development Matters in Saudi Arabian Cities

Mirroring the situation in other regions of the world, the dominant matter with regard to city growth in Saudi Arabia in current times has unquestionably been the fast-paced urbanization and the inclination toward and urgency of sustainable urban development. As illustrated in Table 2.1, all major Saudi Arabian cities are characterized by rapid urbanization. City managers are battling to create spatially cohesive, ecologically balanced, communally all-encompassing, and economically sustainable cities (Bibri et al., 2020; Hallegatte et al., 2020; United Nations, 2018; World Bank, 2020). In describing this situation, the urbanization literature highlights the increasing importance of rapid urban growth as a prerequisite for modernization, development, development programs, and other types of cultural development (Lange, 2017) in Saudi Arabian cities, and the willingness to pass on the lessons learned from the most urbanized regions to less urbanized ones (Heshmati & Rashidghalam, 2020).

The sustainable development literature on Saudi Arabia, on the other hand, places a fundamental focus on negative characterization of urbanization-related issues, including slum growth, social violence, hunger, pollution, congested roads, and shanty towns (Oloke et al., 2021). This, of course, is not limited to Saudi Arabian

Table 2.1 Saudi Arabia's urbanization patterns and the five largest urban agglomeration patterns in Saudi Arabia from 2015 to 2035 (UN-Habitat, 2019)

Year ('000)	2015 ('000)	2020 ('000)	2025 ('000)	2030 ('000)	2035 ('000)
Total population	31,557	34,710	37,290	39,480	41,317
Urban population	26,249	29,256	31,843	34,143	36,170
Level of urbanization (%)	83.2	84.3	85.4	86.5	87.5
Five largest urban agglomerations					
Riyadh	6218	7231	7953	8547	9058
Jeddah	4035	4610	5022	5388	5710
Makkah	1796	2042	2219	2379	2521
Al-Madinah	1299	1489	1625	1744	1848
Dammam	1080	1253	1376	1478	1566
Total population of the five largest agglomerations	14,428	16,625	18,195	19,536	20,703
Five largest agglomerations as % of urban population	54.97	56.82	57.14	57.22	57.24
Five largest agglomerations as % of total population	45.72	47.90	48.79	49.48	50.12

cities, as recent evidence indicates that urban marginalization and deprivation have increased as a result of rapid urbanization occurring in cities in the Global South and the formation of metropolises (Cobbinah et al., 2015; Shackleton et al., 2021). However, these factors are directly related to the unsustainable nature of urban development and economic insecurity (Pilehvar, 2020). Following an analysis of certain important facts about Saudi Arabian urbanization and urban development, a dialogue on the potential sustainable development implications that arise from the two intersecting literatures is offered in the following sections.

As one of the most urbanized nations on Earth, with eight out of every ten residents living in a metropolitan area, urbanization in Saudi Arabia, in the view of Abou-Korin and Al-Shihri (2015), has different sustainable development implications. With about 34 million population as of 2021 and a projected annual growth rate of 2.7%, the country is set to have over 37 million inhabitants and over 90% urban residents by 2025 (General Authority for Statistics (GAS), 2022). A major sustainable development challenge of this relates to the rapid rise in the older adult population and a reduction in children and youth and the working-age population. The percentage of people over the age of 65 rose from 2.6% in 1985 to 3.1% in 2015. It is predicted that this rate will similarly increase to 6.6% in 2030, and to 16.7% in 2050 (UN-Habitat, 2019). Within this context, providing for the planning requirements of older adults in terms of transportation, housing, and healthcare is an evolving sustainable development hurdle.

With sociocultural shifts, modernity, and education, particularly regarding the role of women, Saudi Arabian cities are experiencing a trend toward a decreasing younger population as people delay having, or choose not to have, children. Urbanization offers greater opportunities for women and girls. An independent strategic goal in Saudi Vision 2030 is boosting women's employment rates. In this sense, it is unsurprising that numerous key corporate job vacancies have been filled by women. A significant development is the gradual increase in Saudi women's labor market participation, from 12% in 2009 to 17% in 2019. By 2020, this percentage is predicted to reach 25% (Alessa et al., 2022; UN-Habitat, 2019). While an increase in opportunities for women is to be encouraged and supported, sustainable development implications for migration must be considered in order to address the challenge of a decreasing younger population.

Since 2015, Saudi Arabia's population has increased by 2.5% yearly, and in the previous 25 years, it has more than doubled. (see Fig. 2.2). The population is expected to rise by 13.5 million by 2050 (UN-Habitat, 2019). The Saudi population is expected to peak at 46.2 million in 2065 and then drop to 44 million by 2100 (UN-Habitat, 2019).

Imbalanced population concentration is another sustainable development challenge of rapid urbanization, with about 84% of its people residing in urban areas. Saudi Arabia's population is concentrated in its five largest urban agglomerations (see Table 2.1). Given that thriving cities are a powerful catalyst, and prerequisite, for improving access to healthcare, socio-economic prospects, and standards of living for everyone, the high urbanization level is unquestionably desirable. Regional population changes continue despite attempts to facilitate a more balanced development (Table 2.2). From 1992 to 2017, the populations of Makkah, Riyadh, and the Eastern Province have exhibited the fastest urbanization rates, of 88%, 91%, and 93%, respectively, in Saudi Arabia (UN-Habitat, 2019). Rural–urban mobility and international mobility are the main drivers of urbanization in these regions. Urban

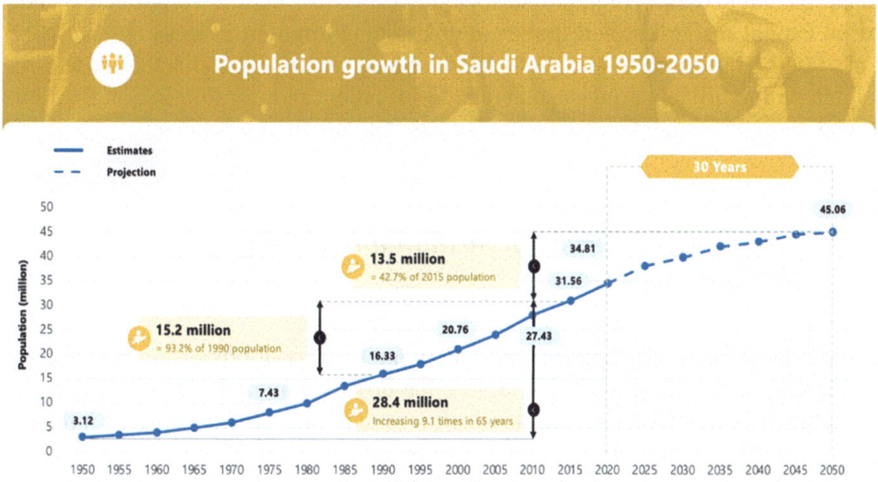

Fig. 2.2 Saudi Arabia's population increase between 1950 and 2050 (Authors' computation using data extracted from UN-Habitat (2019) and General Authority for Statistics (GAS) (2022))

primacy at higher levels has excluded small and medium-sized cities from national development, maintained inequality, and undermined the stability of the state's urban systems (Abou-Korin & Al-Shihri, 2015). In Saudi Arabian cities, this occurs because of growing urban populations and their spatial concentration, economic development, a lopsided investment strategy that prioritizes some urban areas over others, the transformation of rural areas into cities, migration from abroad, and population growth caused by rural-to-urban migration (UN-Habitat, 2019).

The pattern of urbanization in Saudi Arabia raises questions about the adequacy of urban planning as a tool to facilitate sustainable development. Planning at multiple geographic scales reduces urbanization's impact on sustainable development. As argued by Mandeli (2019), urban planning and management are two essential elements for achieving sustainable development and promoting good quality of life, for both permanent and temporary city dwellers. While Saudi Arabia's urban development has been governed by a spatial plan seeking to accomplish an integration between all regions and encourage a spatially equitable population distribution within the nation, its dynamics have unmistakably and selectively favored some significant urban agglomerations, which have become the main recipients of the economic success from its oil income and productivity expansion (UN-Habitat, 2019). Badach et al. (2018) advocate for a careful consideration of varied risks and spatiotemporal variability in urban development strategies across cities to advance efforts at promoting sustainable development across different urban conurbations within the country.

Much of the understanding of sustainable urban development in Saudi Arabia can be derived from the influences of rapid urbanization and urban planning and their implications for land use, land market, leadership at distributed tiers, zoning implementation and tax collection, public spending on infrastructure, corporate expenditure in capital resources, and domestic or capital formation in housing (see Bibri et al., 2020; Hallegatte et al., 2020; World Bank, 2020). This chapter adds to the insights on the urbanization–sustainable development relationship in Saudi Arabia by analyzing the situational barriers of sustainable urban development.

2.4.2 *Situational Barriers to Sustainable Urbanization in Saudi Arabia*

This section focuses on discussing the situational barriers of urbanization to sustainable development. Alshebli (2018) argues that urban pathologies including deprivation, joblessness, underemployment, a decline in real earnings from chronic inflation, and uncontrolled migration are complicated by urbanization. That sustainable development of Saudi Arabian cities is in jeopardy because of urbanization-induced socioeconomic and environmental limitations (Aina et al., 2019; Almulhim & Abubakar, 2021), including urban sprawl, infrastructural difficulties, ecological pollution,

Table 2.2 Saudi Arabian regions: changes in total population (1992–2017) (UN-Habitat, 2019)

Saudi provinces	1992 population	%	2004 population	%	2010 population	%	2017 population	%	Population increase
Al-Bāḥah	332,157	1.96	377,900	1.67	411,888	1.52	491,900	1.49	159,743
Northern Frontier	229,060	1.35	279,971	1.23	320,524	1.18	383,100	1.16	154,040
Al-Jawf	268,228	1.58	361,378	1.60	440,009	1.62	528,400	1.60	260,172
Madinah	1,084,947	6.40	1,512,724	6.67	1,777,933	6.55	2,154,100	6.54	1,069,153
Al-Qaseem	750,979	4.43	1,015,97	4.48	1,215,858	4.48	1,464,800	4.45	713,821
Riyadh	3,834,986	22.63	5,458,273	24.07	6,777,146	24.97	8,276,700	25.12	4,441,714
Eastern Region	2,575,820	15.20	3,360,031	14.82	4,105,780	15.13	4,977,500	15.11	2,401,680
Aseer	1,340,168	7.91	1,687,939	7.44	1,913,392	7.05	2,288,500	6.95	948,332
Ḥā'il	411,284	2.43	526,882	2.32	597,144	2.20	715,400	2.17	304,116
Jīzān	865,961	5.11	1,187,587	5.24	1,365,110	5.03	1,636,600	4.97	770,639
Makkah	4,467,670	26.36	5,797,184	25.56	6,915,006	25.48	8,479,400	25.73	4,011,730
Najrān	300,994	1.78	420,345	1.85	505,652	1.86	607,100	1.84	306,106
Tabouk	486,134	2.87	691,716	3.05	791,535	2.92	946,300	2.87	460,166
Saudi Arabia	16,948,388	100.0	22,678,262	100.0	27,136,977	100.0	32,949,700	100.0	16,001,312

climate change, and insufficient urban amenities. The sustainable development barriers are discussed under three key themes.

2.4.2.1 Challenges to Sustainable Land Use and Planning

The first of these relates to rapid population growth and urban sprawl. By 2050, 45 million people are anticipated to be living in Saudi Arabia's metropolitan areas. As presented in Figs. 2.3 and 2.4, major Saudi Arabian cities have witnessed considerable spatial expansion over the years, highlighting urbanization pressures and the intensity of sustainable-use planning challenges. Low density and urban sprawl are major land-use planning challenges causing major cities to more than triple in size since the 1980s (UN-Habitat, 2019).

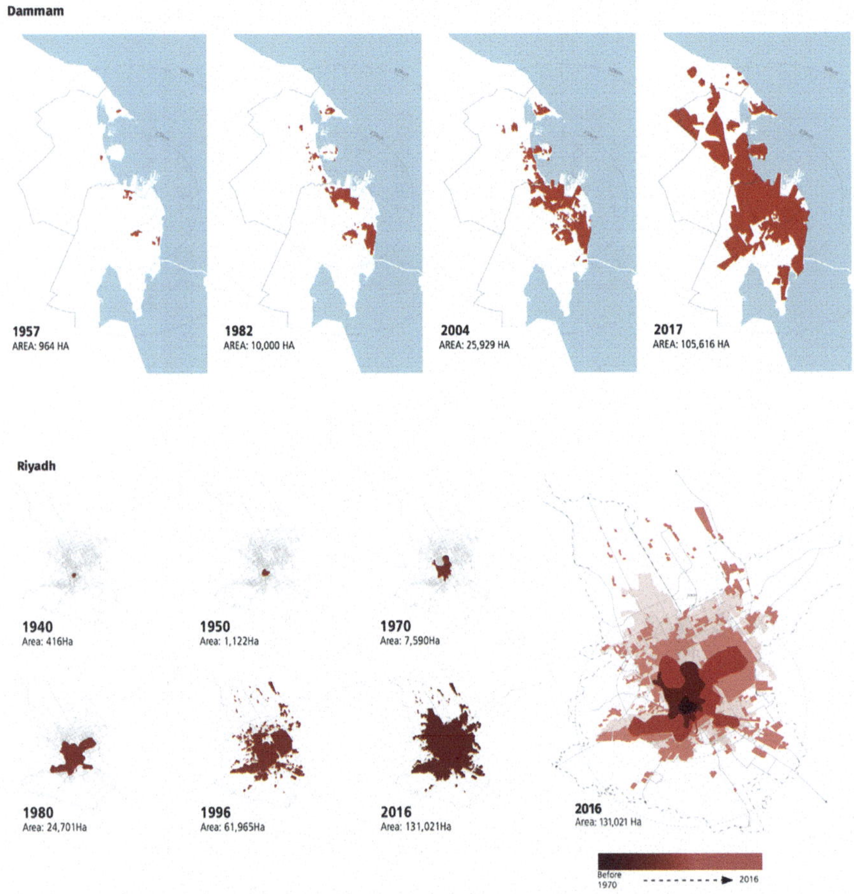

Fig. 2.3 Urban growth stages for Dammam and Riyadh (UN-Habitat, 2019)

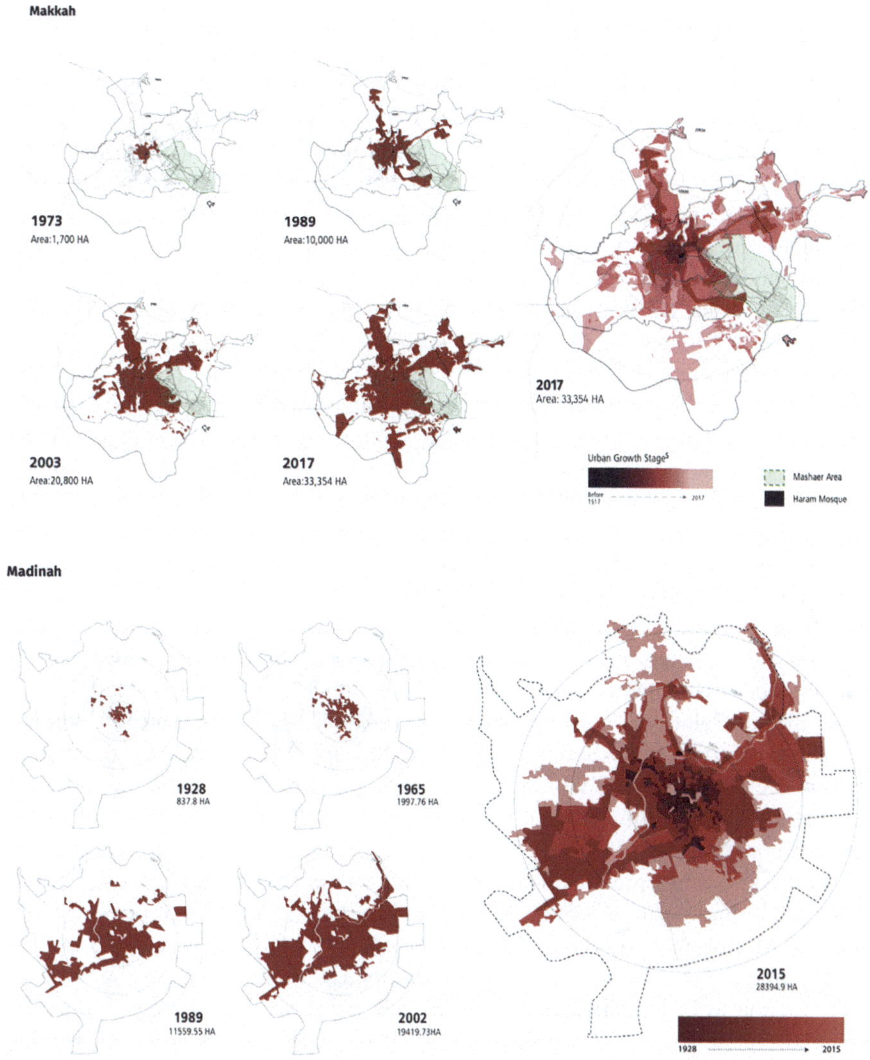

Fig. 2.4 Urban growth stages for Makkah and Madinah (UN-Habitat, 2019)

Densification of unused land should be prioritized over city expansions to sustainably address the continuing increase in Saudi Arabia's urban population. Given the large undeveloped spaces across Saudi Arabian cities, creative urban planning solutions, such as urban aggregation and compact urbanization, are necessary. The primary management strategy used in all significant Saudi cities is expansion of urban boundaries. Huge and unbroken urban aggregates produced by such expansion are unsustainable, posing a formidable challenge to sustainable urban development. A prevalent and urgent issue in all Saudi metropolitan regions is urban area utilization through sprawl. Although these cities are exhibiting rapid population expansion, the physical expanse of urban areas is growing more rapidly, reflecting the rapid consumption of the nation's urban and surrounding land (Alqurashi, 2021; UN-Habitat, 2019).

Land markets remain another challenge. The administrative ability or mechanisms for valuing land are limited. Zoning choices directing urban development cannot be determined due to inefficient planning systems, leading to unfavorable consequences including irregular density patterns, expansion, and informality. For government agencies, Saudi Arabia's rapid urbanization has presented several difficulties. The primary local issues are wide-scale sprawl, inefficient use of environmental assets, and the growing demand for facilities and services.

Complicating matters further is the type of local governance practiced. Saudi Arabia's centralized government continues to dominate urban planning and management. Despite initiatives to decentralize, the central government retains considerable power in local politics, including the authority to terminate regional authorities, dismiss local council members, and determine local tax rates. This situation is occurring despite urban governance being expected to provide a basic framework for non-state actors to participate in governance activities through feedback. Elections for city council members were only recently introduced in the country, and involving the public in politics is still difficult.

Contrary to expectations, though, larger cities (e.g., Dammam, Riyadh) typically exhibit greater citizen involvement in urban development compared to smaller cities because of their relatively cosmopolitan identity. While it is important to acknowledge recent efforts by the Saudi government through Vision 2030 to drastically transform the country into a vibrant society with a booming economy and efficient government (Khan & Iqbal, 2020; UN-Habitat, 2019), citizen involvement in the governance process is still limited.

Solid-waste management is a major planning concern. Municipal solid waste has increased due to, among other things, rapid urbanization and improved living standards (Almulhim & Al-Saidi, 2023). Saudi Arabia produces at least 15 million tons of garbage annually, which is mostly dumped without being treated. Municipal production in Dammam, Jeddah, and Riyadh surpasses 6 million tons annually (Elboughdiri et al., 2022). The government's Vision 2030 acknowledges the need to develop strategies and regulations to enhance sewage treatment as landfills produce 76% of methane leaks. Vision 2030 aims to decrease pollution from trash, particularly plastic materials, and boost the effectiveness of waste management by building extensive recycling projects. However, Saudi Arabian cities must transform their domestic behavior by raising public awareness of waste prevention and reduction, and by fostering consumer attitudes that support environmentally friendly behavior. This shift is difficult because of inadequate infrastructure, scarce resources, and improper urban planning and management systems (see Abdulghaffar & Williams, 2021; Zeng et al., 2022).

2.4.2.2 Socio-spatial Vulnerability Challenges

Rapid urbanization is subjecting the social and spatial aspects of Saudi Arabian cities to increased vulnerability, and there is immense pressure on infrastructure and essential services. As a country of low population density, the Saudi government has problems providing basic public amenities and services to the dispersed residential neighborhoods in its growing cities. Aljoufie (2021) notes that many houses in the nation's expanding conurbations do not have access to basic essential services such as roads, water, and health. Furthermore, provision of basic amenities (e.g., power, water, and sanitation) has become a financial burden for government due to uncoordinated sprawling of cities (Alqarni, 2017). Rapid urbanization amidst low population density patterns development contributes to unsustainable development and inefficiency by raising energy and transportation costs, making it more expensive to manage public spaces and infrastructure, lessening the benefits of agglomeration, and impairing productivity in cities. Local governments do not possess the means, access (to state counsel), or power to address this issue, despite considerable recent infrastructural enhancements that have improved key delivery of services. UN-Habitat (2016b) called for increased citizen involvement in city governance to improve planning, budgeting, management, and monitoring of service delivery.

A major evolving vulnerability challenge relates to unemployment in Saudi Arabian cities. Annually, approximately 400,000 Saudis enter the workforce. In the second quarter of 2021, 15.4% of people were unemployed. This unemployment rate is higher than other countries, despite the recent academic success of Saudi youth (see Al-Jarf, 2021). Providing unemployment assistance to all Saudis between the ages of 20 and 60 years and constantly seeking employment coverage is a strategy suggested by NTP 2020 to improve the country's workforce participation (Grand & Wolff, 2020). The number of job seekers is increasing in both traditional and modern sectors of the urban economy, a key effect of accelerated urban expansion. Unemployment prevents people from enjoying economic independence and influences their attitudes toward protecting the environment and forming strong social bonds. The homogeneous activities dominating the countryside (agriculture) coupled with rural-to-urban migration are shifting the unemployment rate from rural to metropolitan areas (Abubakar & Aina, 2018).

Rising inequality compounds the situation further. The average Gini coefficient in Saudi Arabian cities is about 45.9, which is higher than in the U.S. However, inequality in Saudi Arabia is increasing. Although the government withholds official numbers, private organizations estimate that approximately 20% of native Saudis are poor (Stoffel, 2017). Saudi Arabian cities with a high concentration of slums and informal settlements experience increased poverty, informality, and inequality. National urban regulations have not yet facilitated multi-level management through well-defined regulatory frameworks, founded on the tenets of decentralization and subsidiarity, and through sufficient government management of economic resources (see UN-Habitat, 2019).

Transport systems, traffic congestion, and accidents remain another important challenge. There are 219 cars per 1000 inhabitants, reflecting the deeply ingrained dependence on private vehicle ownership (Dano & Alqahtany, 2019). This has been achieved through land-use policies and layout plans that allow for highways and abundant parking along the road, and domestic increases in suburban automobile use due to several tracts of inexpensive open regions on the outskirts of metropolitan centers. Residents have to travel long distances for work, and public transportation is unreliable (due to the rise in the number of vehicles) and inadequate. As expected, traffic congestion has become commonplace in cities, impacting on productivity and urban sustainability (see Alotaibi & Potoglou, 2018).

In addition, inadequate affordable housing is a key factor contributing to Saudi Arabia's housing crisis. Despite the abundance of high-rent accommodation in most areas, low- and middle-income families are facing a chronic housing crisis. There is an urgent need for housing that is accessible to all citizens and fits their socio-economic needs. Since 1974, despite the difficulties of rapid urbanization, the supply of inexpensive, hygienic accommodation has been a top priority and a pillar of Saudi Arabia's urban policy (Aljoufie & Tiwari, 2020). For example, despite the Ministry of Housing being responsible for building residences, the government has offered free land and interest-free mortgages via the Real Estate Development Fund, which is the primary financier of housing. Between 2008 and 2013, the government increased the yearly cost of the fund from USD 13,500 million to USD 62,100 million (Ministry of Municipal, Rural Affairs & Housing, 2020), yet despite this, cities in Saudi Arabia face severe housing shortages (Ramady, 2021; UN-Habitat, 2019). The prevalence of uninhabited "white lands" has significantly contributed to an increasing housing scarcity, especially for young people as land speculation intensifies. In addition, a number of housing units are vacant because of the widening disparity between asking prices and household income for low- and middle-income families. This group accounts for about 80% of the unfulfilled demand. It is hoped that the implementation of the objectives of the NTP and Vision 2030 will be a turning point for housing supply as they target the property market (see UN-Habitat, 2019).

2.4.2.3 Urban Environmental Challenges

Over the past decade, the country has made efforts to conserve and utilize natural resources sustainably for increasing wealth, preserving human rights, securing livelihoods, and ensuring reasonable living standards (Almulhim, 2022). However, rapid urbanization poses critical environmental challenges such as encroachment, habitat loss, and acceleration of desertification. Furthermore, water scarcity is a pertinent obstacle to improving living standards and ensuring progress toward sustainable development in the country (Almulhim & Aina, 2022). Increased temperatures and protracted droughts attributed to climate change exacerbate the water scarcity issue by imposing additional demand on limited water supplies. This situation is heightened by urbanization, which is increasing water supply shortages nationwide. Moreover, environmental challenges in metropolitan areas severely impact the country's economy and affect citizens' mental health and overall health.

Climate change has become a fundamental challenge. The country's climate is among the most dangerous worldwide, with 108 μg of airborne fine particulates per m^3 (PM 2.5), surpassing the WHO's tolerable guideline of 10 $\mu g/m^3$ of PM 2.5 (Nayebare et al., 2022; World Air Quality Report, 2021). As cities produce over 70% of carbon emissions (World Bank, 2022), urban areas are essential for the mitigation of and adaptation to climate change. In the past three decades, 14 natural disasters have been documented in Saudi Arabia, affecting close to 30,000 lives and causing an estimated USD 450 million in economic losses (UN-Habitat, 2019). The dangers of global warming include increased incidence and harshness of heatwaves; an increase in dust storms, possibly impacting tourism and agriculture; changes in the frequency, type, and intensity of rainfall; and extreme or unusual flooding (UN-Habitat, 2019). The integration of disaster risk-reduction initiatives with ongoing climate change adaptation activities can support urban sustainable development initiatives. Saudi cities must re-evaluate their plans for spatial growth and incorporate global warming factors into planning and design of their drainage systems, building regulations, and road systems, and direct more efforts toward expanding public knowledge on climate change (Almulhim, 2021; UN-Habitat, 2018, 2019). The Sendai Framework for Disaster Risk Reduction (2015–2030) entails considering age and gender when designing infrastructure and using ecosystem-based methods (UN-Habitat, 2019). Managing environmental changes requires multi-level governance architecture.

The above analysis raises the question of how to address the limitations of urbanization in Saudi Arabia and engender sustainable urbanization. Enabling sustainable urbanization remains a priority for major international sustainable development agendas, including the SDG 11, and the Sendai Framework for Disaster Risk Management. How can cities in Saudi Arabia enable sustainable urbanization to address the challenges of urban growth and promote inclusive, sustainable, and resilient urban spaces? The next section analyzes the enablers of sustainable development in Saudi Arabia.

2.4.3 Enabling Sustainable Urbanization in Saudi Arabia

The ultimate objective for all cities is livability and functionality. Cities must be inclusive, sustainable, competitive, and resilient to be livable. We propose three pillars to enable sustainable urbanization in Saudi Arabian cities: (a) resilient cities—to understand community capacity to respond to stresses in term of social, economic, and natural shocks, (b) inclusive cities—to appreciate all segments of society and all groups, including the vulnerable and the marginalized, in their markets, services, and development, and (c) sustainable and competitive cities—to understand how cities can sustain productivity and create jobs for urban residents, with a focus on the tripartite aspects of sustainable development (i.e., environmental, social, and economic dimensions). We argue that to reduce poverty and increase shared prosperity across local, regional, and national levels in Saudi Arabia, it is imperative to build inclusive, resilient, competitive, and sustainable cities and to realize the objects of the Sustainable Development Goals by 2030. As illustrated in Fig. 2.5, we use SDG 11, which addresses sustainable communities and societies, to put forward a sustainable urbanization agenda for Saudi Arabia, and by implications other Arabian cities. First, we categorized the targets of SDG 11 into three, based on the urbanization narrative in Saudi Arabia (see Sect. 2.4). We further translated the urbanization challenges and pressures into opportunities and benefits and aligned them with the appropriate targets of SDG 11, ensuring that implementation of any of the targets will lead to realization of the urbanization opportunities and benefits. Finally, we discussed these three pillars of sustainable development within the global urbanization literature to demonstrate their relevance, particular in Saudi Arabian cities. The following paragraphs provide explanations for our three-pillar framework of sustainable development.

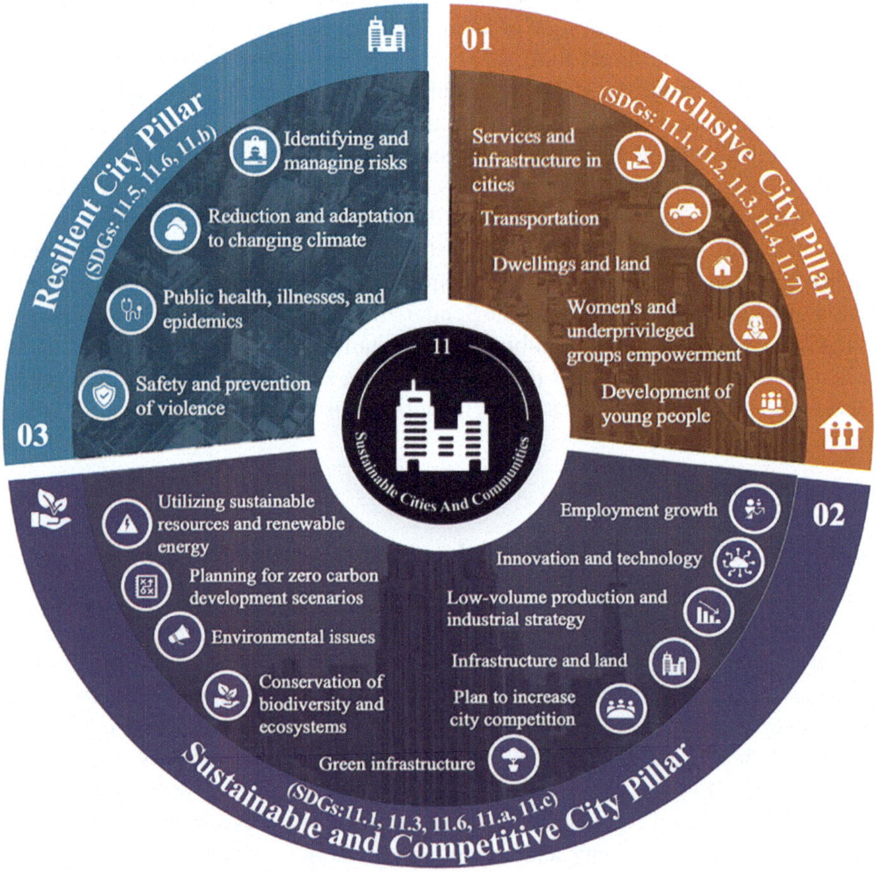

Fig. 2.5 The three key pillars proposed for sustainable urbanization (illustration created by the authors)

The inclusive city pillar aims to address the structural gaps between the powerful and disenfranchised in an inclusive manner. Each resident is valued in an inclusive city, which offers opportunities to create an urban life through participatory planning, open budgeting, and fair political representation. Increased financial mobility and mutual wealth are made possible, and along with creating employment, this pillar offers opportunities for everyone to achieve stable and sustainable living by improving the urban populations' skills and training. The equitable delivery of services and protection of vulnerable communities in urban projects and development are essential elements of this pillar. Access to essential amenities, accommodation, healthcare, and education is a prerequisite for this pillar; indeed, authentic inclusion is predicated on a city's capacity to incorporate the requirements of all its demographic groups. This will ensure protection of residents' civil and human rights through robust and fair governance structures. Moreover, measures for accountability and transparency should be implemented to guard against corruption and the sway of powerful interest groups.

With *the sustainable and competitive pillar*, we argue that well-managed cities should exhibit significant societal well-being and economic progress while using fewer resources and generating less carbon and other emissions. By implication, sustainable cities should lead their countries and regions toward environmental protection via a systems-based transformational approach, an approach that takes advantage of economies of size and scope and supports rapid cultural and organizational creativity, to improve its economy, increase productivity, and boost all its residents' incomes. We argue that it is only when these occur that a competitive city will optimally utilize its assets through sound policymaking and governance (a key component of competitiveness) together with equity and poverty reduction.

Agglomeration (the critical collating of facilities, economic factors [including labor and raw materials], companies, and industries) is one of the essential elements for the sustainable and competitive pillar. The principal benefit of conurbations is their agglomeration of businesses, which promotes industry expansion and high levels of competition. Decisive action, to implement policies and initiatives enhancing financial growth, urban regeneration, startups, and skill building into practice, is another central component of competitiveness. When policy initiatives are coordinated to improve fairness, general development, resilience, and financial activities, a city sustains the benefits from its competitiveness.

The resilient city pillar involves building resilient cities to deliver sustainable outcomes. We contend that a resilient city should be prepared to adjust to recurring stressors and sudden physical, societal, and economic disruptions resulting from both climate change and rapid urbanization. To ensure the preparedness of urban populations to recuperate from their problems, this pillar focuses on building resilience in cities. Applying this pillar, we advocate for risks to be thoroughly identified and evaluated by city managers, emphasizing potential repercussions for disadvantaged groups via an open and participatory process. All key stakeholders, particularly those affected, must be involved and empowered to accurately estimate multi-hazard threats. We therefore suggest that long-term urban and regional planning must incorporate a resilience strategy, which can maintain quality and prosperity, despite uncertain external conditions. This will lead to the production of resilient cities that exhibit solid vision, cross-sector collaboration, and substantial strategic choices, precipitating efficient urban management, particularly in times of crises.

2.4.3.1 What the Pillars Mean for Cities in Saudi Arabia and Across the Global South

Urbanization has become a defining feature of modern urban life and is central to how cities are planned, developed, and managed. Beyond its conventional foundational conceptions, practice of urban planning across the Global South attempts to respond to and address the many complications associated with urbanization (see Cobbinah et al., 2015; Elmqvist et al., 2021). The analysis in this chapter indicates that in Saudi Arabia, urbanization can favorably impact socio-economic development if the regulations and frameworks aiming to achieve development goals are met. However, urbanization debates in the country suggest enormous limitations that must be addressed. We argue, through the development of the three-pillar sustainable development analysis in this chapter, that a cohesive unified policy and coordinated approach transcending governmental entities, both vertically, from local to national, and horizontally, throughout ministries, is required to capitalize on urbanization and to create superior urban opportunities. In urban studies, comprehensive national urbanization frameworks that fundamentally establish the groundwork and frame future guidelines to leverage the advantages of urbanization, for the advancement of an entire nation are imperative (see Spicer, 2021; UN-Habitat, 2014). Drawing scenarios from urbanization management experiences via the three-pillar sustainable development analysis, we show that a coherent regional-spatial strategy, public revenue options, and activities by all decision makers can be coordinated through national urban policy, to prioritize socio-economically and ecologically sustainable urban development areas in Saudi Arabian cities.

Many scholars have presented different cases from numerous urban regions across the Global North and South emphasizing the complexity of the urbanization phenomenon and its driving forces (Abou El-Magd et al., 2022; Adelina et al., 2020; Aina et al., 2019). Our three-pillar sustainable development analysis demonstrates that a diagnosis of the factors causing the urban transition should be included in national urbanization management frameworks, along with information on the functions of the general and specific cities within the national environment. Within this context, urban regulatory areas in Saudi Arabian cities should be considered to advance preferred trends of urban spatial development, guarantee satisfactory rules for land ownership and legal protections, provide comprehensive building and zoning standards, and differentiate between the roles and functions of national and regional local authorities.

Similarly, our three-pillar sustainable development analysis indicates that housing policy must contain provisions to fulfill the predicted increasing demands of unit accessibility and price, ensuring cooperation between the public and corporate elements. Across Saudi Arabian cities, a national legislative framework on the production, distribution, and financing of urban and regional services and facilities, particularly, power, water, hygiene, telecommunication, and mobility, is essential. Considering that cities in the Global South have emerged as new centers of urbanization yet highly vulnerable to its negative implications, urbanization management has become ever more pressing (Abou El-Magd et al., 2022; Cobbinah et al., 2015; Elmqvist et al., 2021). Our three-pillar sustainable development analysis explains that to maximize Saudi Arabian cities' unique quality within the regional supply chain and promote sustainable prospects from urbanization, particularly for women and youngsters, urban economic strategy is imperative. As such, national urban policy must consider the dynamics and interconnections among principal cities and medium-sized towns, including villages with lesser populations (see Dianat et al., 2022; UN-Habitat, 2015) in order to set sustainable urban management agendas in motion.

2.5 Discussion and Conclusion: The Agency of Sustainable Urbanization in Saudi Arabia

The main objective of this chapter has been to understand how the dynamics of urbanization, as described commonly worldwide and particularly in the Global South, have impinged on sustainable urban development and planning in Saudi Arabian cities. Reflecting a more conventional pattern in global literature, findings from this research show that urbanization in Saudi Arabian cities is associated with more costs (challenges) than benefits (opportunities), a situation that is threatening sustainable urban development. However, urbanization in Saudi Arabian cities is distinctive in many ways in the context of sustainable development and urban planning, and is worth reflecting on. Findings from this research indicate that urbanization in the context of sustainable development has socio-spatial, land-use planning, and environmental vulnerabilities. While the socio-spatial vulnerability predisposes urban citizens to poverty, housing and related crises, land-use planning vulnerability is contributing to unsustainable land use and difficulties with service provision. The environmental limitations focus on encroachment on and destruction of natural areas as well as climate-related impacts, which are generating urban health-related concerns.

The introduction, availability, and interaction of globally, regionally, nationally, and locally oriented sustainable development policy frameworks such as SDGs, New Urban Agenda, the Saudi government's Vision 2030, and NTP are driving interest and commitment toward addressing rapid urbanization in Saudi Arabian cities. However, not much has been achieved in this regard, as the negative implications of urbanization dominate in Saudi Arabian cities. Adding to the globally reported outcomes of urbanization such as unemployment, housing crises, traffic congestion and uncontrolled urban sprawl, findings from this chapter show a gradual improvement of women's role in the economy in more urbanized cosmopolitan areas compared to less urbanized ones. Unfortunately, improved female participation in leadership roles and economic activities is occurring in the absence of widespread citizen involvement in urban governance.

While urbanization has been widely discussed from many perspectives (Alshuwaikhat & Mohammed, 2017; Cobbinah et al., 2015; Kisiała & Rącka, 2021; Ochoa et al., 2018), this chapter shows that its impacts are severe in the Arab world, particularly in Saudi Arabian cities where the harsh realities of climate change conflate urbanization outcomes. Despite this, limited studies and meaningful direction are offered to city managers, governments, and policy makers in Saudi Arabia regarding an encompassing agenda required to address this growing occurrence. Similarly, the activities of politics (e.g., government's role) that are supposed to promote sustainable development of cities are in many ways frustrating and weakening urban planning attempts, as they consciously or accidentally interfere in and dictate planning actions. Complicating matters further are the vulnerabilities of city residents with their limited opportunity to engage and participate in planning and governance efforts, which is contributing to unsustainable patterns of urban development, illustrated through inadequate infrastructure and essential services, unfavorable resource allocation, and uncoordinated urban development.

Given the ongoing pattern of urbanization in Saudi Arabian cities and the diversity of intertwined issues, it is clear that it will continue to be tough for city managers, governments, and policy makers to adequately address this phenomenon without considerable involvement of urban citizens and an agenda that recognizes the three pillars of inclusiveness, sustainability and competitiveness, and resilience. Non-involvement of urban citizens in this manner has mostly led to catastrophes in terms of clearly paying attention to the issues that are central to citizens and their capacity to manage urbanization. These include acknowledging the value of citizen involvement and adapting urban planning systems to accommodate them, as well as recognizing inclusiveness, resilience and sustainability in the process of urbanization management; equipping and empowering city planners and managers to be independent in working with city citizens to direct, plan, and manage rapid urbanization and its related consequences; and encouraging and according priority to research on urbanization–sustainability relationships.

The findings presented in this chapter show that urbanization should be considered not as a modest urban development issue, but rather as a multifaceted sustainable development challenge underpinned by the production of global, regional, national, and local factors that frame and shape the future of many actors, particularly vulnerable urban citizens, and cities. Addressing urbanization in Saudi Arabian cities demands more than institutional responses or government policy direction. The process of managing urbanization should aim to demystify the positionality of urban citizens using the pillars of inclusiveness, sustainability and competitiveness, and resilience. It is only when city planners and managers, governments and policy makers recognize that the many urbanization challenges need to be addressed via an improved structure built on resilience, sustainability, and inclusiveness, and defined by participatory processes involving all key actors, can efforts toward sustainable urbanization come to fruition.

In sum, this chapter makes three important contributions to urban studies and planning literature. First, it is among the first few works to analyze the urbanization–sustainable development relationship in an Arabian context. While available literature (Kiamba, 2012; Rana & Parves, 2011) indicates that little is known about the urbanization–sustainable development nexus in the Arabian context despite the mushrooming of literature on the subject in other Asian and African countries, this chapter contributes to addressing this gap by establishing the positionality of the urbanization–sustainable development debate in Saudi Arabia. Secondly, the chapter offers two analytical perspectives on urbanization: population growth and spatial extent. Previous literature (e.g., Bibri et al., 2020; Cohen, 2006; Hallegatte et al., 2020) mostly focuses on either population growth or spatial extent in the urbanization debates. By contrast, this chapter addresses both debates from a region experiencing unprecedent urbanization levels. Finally, this chapter is the first to explore enablers for sustainable urbanization in the Saudi Arabian context, highlighting the embeddedness of socio-spatial factors and land-use planning systems in the production of environmental vulnerabilities. Previous studies frequently identify the challenges or limitations of urbanization, but this chapter goes beyond identification of problems to development of solutions.

Despite these contributions, there was one major limitation. The part of the analysis focusing on secondary data was predominantly generated from Saudi government and planning institutions documents (e.g., Saudi Cities Report, 2019). There is the possibility that this situation could have practically restricted the perspectives, experiences, and realities of everyday residents in relation to urbanization-related issues, and championed the agenda of the Saudi government and urban planning agencies. Despite this limitation, the use of critical literature review employing both secondary data and scholarly sources provided a framework to address the limitation.

Acknowledgements This chapter is based on the following previously published article and is reproduced with permission from Elsevier: Almulhim, A.I., & Cobbinah, P.B. (2023). *Can rapid urbanization be sustainable? The case of Saudi Arabian cities.* Habitat International, 139, 102884.

References

Abdulghaffar, N. A., & Williams, I. D. (2021). Development of sustainable waste management in higher education institutions. *AIMS Environmental Science, 8*(3), 238–254. https://doi.org/10.3934/environsci.2021016

Abou El-Magd, I., Attwa, M., El Bastawesy, M., Gad, A., Henaish, A., & Zamzam, S. (2022). Qualitative and quantitative characterization of municipal waste in uncontrolled dumpsites and landfills using integrated remote sensing, geological and geophysical data: A case study. *Sustainability, 14*(8), 4539. https://doi.org/10.3390/su14084539

Abou-Korin, A. A., & Al-Shihri, F. S. (2015). Rapid urbanization and sustainability in Saudi Arabia: The case of Dammam metropolitan area. *Journal of Sustainable Development, 8*(9), 52. https://doi.org/10.5539/jsd.v8n9p52

Abubakar, I. R., & Aina, Y. A. (2018). Achieving sustainable cities in Saudi Arabia: Juggling the competing urbanization challenges. In Management Association, I. (Ed.), *E-planning and collaboration: Concepts, methodologies, tools, and applications* (pp. 234–255). IGI Global. https://doi.org/10.4018/978-1-5225-5646-6.ch011

Abubakar, I. R., & Dano, U. L. (2020). Sustainable urban planning strategies for mitigating climate change in Saudi Arabia. *Environment, Development and Sustainability, 22*(6), 5129–5152. https://doi.org/10.1007/s10668-019-00417-1

Addas, A., & Maghrabi, A. (2021). Role of urban greening strategies for environmental sustainability—A review and assessment in the context of Saudi Arabian megacities. *Sustainability, 13*(11), 6457. https://doi.org/10.3390/su13116457

Adelina, C., Archer, D., Romanus, O. J., & Opiyo, O. (2020). *Governing sustainability in secondary cities of the Global South.* Stockholm Environment Institute.

Ahani, S., & Dadashpoor, H. (2021). A review of domains, approaches, methods and indicators in peri-urbanization literature. *Habitat International, 114*, Article 102387. https://doi.org/10.1016/j.habitatint.2021.102387

Aina, Y. A., Wafer, A., Ahmed, F., & Alshuwaikhat, H. M. (2019). Top-down sustainable urban development? Urban governance transformation in Saudi Arabia. *Cities, 90*, 272–281. https://doi.org/10.1016/j.cities.2019.03.003

Alderton, A., Higgs, C., Davern, M., Butterworth, I., Correia, J., Nitvimol, K., & Badland, H. (2021). Measuring and monitoring liveability in a low-to-middle income country: A proof-of-concept for Bangkok, Thailand and lessons from an international partnership. *Cities & Health, 5*(3), 320–328. https://doi.org/10.1080/23748834.2020.1813537

Alessa, N. A., Shalhoob, H. S., & Almugarry, H. A. (2022). Saudi women's economic empowerment in light of Saudi Vision 2030: Perception, challenges and opportunities. *Journal of Educational and Social Research, 12*(1), 316–316. https://doi.org/10.36941/jesr-2022-0025

Aljaddani, A. H., Song, X.-P., & Zhu, Z. (2022). Characterizing the patterns and trends of urban growth in Saudi Arabia's 13 Capital Cities using a Landsat Time Series. *Remote Sensing, 14*(10), 2382. https://doi.org/10.3390/rs14102382

Al-Jarf, R. (2021, October). The pandemic job market from the perspective of female languages and translation college graduates in Saudi Arabia: Crisis or opportunity. In A. Rocha & E. Isaeva (Eds.), *International Perm forum: Science and global challenges of the 21st century* (pp. 720–733). Springer, Cham.

Aljoufie, M. (2021). The impact assessment of increasing population density on Jeddah road transportation using spatial-temporal analysis. *Sustainability, 13*(3), 1455. https://doi.org/10.3390/su13031455

Aljoufie, M., & Tiwari, A. (2020). Exploring housing and transportation affordability in Jeddah. *Housing Policy Debate*, 1–27. https://doi.org/10.1080/10511482.2020.1815070

Almulhim, A. I. (2021). Public knowledge and perception of climate change and global warming in the context of environmental challenges and policies in Saudi Arabia. *WIT Transactions on Ecology and the Environment, 253*, 577–589. https://www.witpress.com/Secure/elibrary/papers/SC21/SC21047FU1.pdf

Almulhim, A. I. (2022). Understanding public awareness and attitudes toward renewable energy resources in Saudi Arabia. *Renewable Energy, 192*, 572–582. https://doi.org/10.1016/j.renene.2022.04.122

Almulhim, A. I., & Abubakar, I. R. (2021). Understanding public environmental awareness and attitudes toward circular economy transition in Saudi Arabia. *Sustainability, 13*(18), 10157. https://doi.org/10.3390/su131810157

Almulhim, A. I., & Aina, Y. A. (2022). Understanding household water-use behavior and consumption patterns during COVID-19 lockdown in Saudi Arabia. *Water, 14*(3), 314. https://doi.org/10.3390/w14030314

Almulhim, A. I., & Al-Saidi, M. (2023). Circular economy and the resource nexus: Realignment and progress towards sustainable development in Saudi Arabia. *Environmental Development, 46*, Article 100851. https://doi.org/10.1016/j.envdev.2023.100851

Almulhim, A. I., Bibri, S. E., Sharifi, A., Ahmad, S., & Almatar, K. M. (2022). Emerging trends and knowledge structures of urbanization and environmental sustainability: A regional perspective. *Sustainability, 14*(20), 13195. https://doi.org/10.3390/su142013195

Almulhim, A. I., & Cobbinah, P. B. (2022). Urbanization–environment conundrum: An invitation to sustainable development in Saudi Arabian cities. *International Journal of Sustainable Development & World Ecology*, 1–15. https://doi.org/10.1080/13504509.2022.2152199

Alotaibi, O., & Potoglou, D. (2018). Introducing public transport and relevant strategies in Riyadh City, Saudi Arabia: A stakeholders' perspective. *Urban, Planning and Transport Research, 6*(1), 35–53. https://doi.org/10.1080/21650020.2018.1463867

Alqahtany, A., & Aravindakshan, S. (2021). Urbanization in Saudi Arabia and sustainability challenges of cities and heritage sites: Heuristical insights. *Journal of Cultural Heritage Management and Sustainable Development* (ahead of print). https://doi.org/10.1108/JCHMSD-07-2020-0108

Alqarni, M. H. M. (2017). *Developing a framework to improve the implementation of geospatial technology in the planning and delivery of infrastructure for residential areas in Saudi Arabia: A case study of Riyadh City*. University of Northumbria.

Alshebli, A. (2018). *Towards making urban planning practices more effective amid rapid urban growth in Riyadh-Saudi Arabia* (Doctoral dissertation), University of Birmingham.

Alshuwaikhat, H. M., & Mohammed, I. (2017). Sustainability matters in national development visions: Evidence from Saudi Arabia's Vision for 2030. *Sustainability, 9*(3), 408. https://doi.org/10.3390/su9030408

Al-Wabel, M. I., Sallam, A., Ahmad, M., Elanazi, K., & Usman, A. R. (2020). Extent of climate change in Saudi Arabia and its impacts on agriculture: A case study from Qassim region. In *Environment, climate, plant and vegetation growth* (pp. 635–657). Springer. https://doi.org/10.1007/978-3-030-49732-3_25

Alqurashi, A. F. (2021). Quantification of urban patterns and processes through space and time using remote sensing data: A comparative study between three Saudi Arabian cities. *Sustainability, 13*(22), 12615. https://doi.org/10.3390/su132212615

Andreola Serraglio, D., Sivini Ferreira, H., & Robinson, N. A. (2019). Climate-induced migration and resilient cities: A new urban agenda for sustainable development. *Revista Seqüência, 42*(83), 10–46.

Asadzadeh, A., Kötter, T., Fekete, A., Moghadas, M., Alizadeh, M., Zebardast, E., Weiss, D., Basirat, M., & Hutter, G. (2022). Urbanization, migration, and the challenges of resilience thinking in urban planning: Insights from two contrasting planning systems in Germany and Iran. *Cities, 125*, Article 103642. https://doi.org/10.1016/j.cities.2022.103642

Badach, J., Kolasińska, P., Paciorek, M., Wojnowski, W., Dymerski, T., Gębicki, J., Dymnicka, M., & Namieśnik, J. (2018). A case study of odour nuisance evaluation in the context of integrated urban planning. *Journal of Environmental Management, 213*, 417–424. https://doi.org/10.1016/j.jenvman.2018.02.086

Bai, X., McPhearson, T., Cleugh, H., Nagendra, H., Tong, X., Zhu, T., & Zhu, Y. G. (2017). Linking urbanization and the environment: Conceptual and empirical advances. *Annual Review of Environment and Resources, 42*(1), 215–240. https://doi.org/10.1146/annurev-environ-102016-061128

Bibri, S. E., Krogstie, J., & Kärrholm, M. (2020). Compact city planning and development: Emerging practices and strategies for achieving the goals of sustainability. *Developments in the Built Environment, 4*, Article 100021. https://doi.org/10.1016/j.dibe.2020.100021

Camhis, M. (2006). Sustainable development and urbanization. In M. Keiner (Ed.), *The future of sustainability* (pp. 69–98). Springer.

Carruthers, D. (2001). From opposition to orthodoxy: The remaking of sustainable development. *Journal of Third World Studies, 18*(2), 93–112.

Christley, E., Ljungberg, H., Ackom, E., & Nerini, F. F. (2021). Sustainable energy for slums? Using the sustainable development goals to guide energy access efforts in a Kenyan informal settlement. *Energy Research & Social Science, 79*, Article 102176. https://doi.org/10.1016/j.erss.2021.102176

CIA. (2019). *The world factbook-South America.* Central Intelligence Agency (CIA).

Cilliers, S. S., Breed, C. A., Cilliers, E. J., & Lategan, L. G. (2021). Urban ecological planning and design in the Global South. In *Urban ecology in the global south* (pp. 365–401). Springer. https://doi.org/10.1007/978-3-030-67650-6

Cobbinah, P. B., Black, R., & Thwaites, R. (2011). Reflections on six decades of the concept of development: Evaluation and future research. *Journal of Sustainable Development in Africa, 13*(7), 134–149.

Cobbinah, P. B., Erdiaw-Kwasie, M. O., & Amoateng, P. (2015). Rethinking sustainable development within the framework of poverty and urbanisation in developing countries. *Environmental Development, 13*, 18–32.

Cobbinah, P. B., & Finn, B. M. (2022). Planning and climate change in African cities: Informal urbanization and "just" urban transformations. *Journal of Planning Literature.* https://www.x-mol.net/paper/article/1577083520946855936

Cohen, B. (2006). Urbanization in developing countries: Current trends, future projections, and key challenges for sustainability. *Technology in Society, 28*(1–2), 63–80.

Dano, U. L., & Alqahtany, A. M. (2019). Issues undermining public transport utilization in Dammam city, Saudi Arabia: An expert-based analysis. *Journal of Sustainability Science and Management, 14*(2), 155–169.

Deb, S., & Okulicz-Kozaryn, A. (2023). Exploring the association of urbanisation and subjective well-being in India. *Cities, 132*, Article 104068. https://doi.org/10.1016/j.cities.2022.104068

Dianat, H., Wilkinson, S., Williams, P., & Khatibi, H. (2022). Choosing a holistic urban resilience assessment tool. *International Journal of Disaster Risk Reduction, 71*, Article 102789. https://doi.org/10.1016/j.ijdrr.2022.102789

Du Pisani, J. A. (2006). Sustainable development–Historical roots of the concept. *Environmental Sciences, 3*(2), 83–96.

Elboughdiri, N., Khan, M. I., & Missaoui, T. (2022). Estimating the energy recovery potential from municipal solid waste in Ha'il region (Kingdom of Saudi Arabia). *Yanbu Journal of Engineering and Science, 19*(1), 11–17. https://doi.org/10.53370/001c.31893

Elimam, H. (2022). Environmental problems & development sustainability in light of the Kingdom's 2030 Vision: Opportunities & challenges. *International Journal of Education and Social Science, 9*(1), 2410–5171.

Elmqvist, T., Andersson, E., McPhearson, T., Bai, X., Bettencourt, L., Brondizio, E., Colding, J., Daily, G., Folke, C., Grimm, N. B., Haase, D., Ospina, D., Parnell, S., Polasky, S., Seto, K. C., & Van Der Leeuw, S. (2021). Urbanization in and for the Anthropocene. *npj Urban Sustainability, 1*(1), 1–6. https://doi.org/10.1038/s42949-021-00018-w

Fukuda-Parr, S., & Muchhala, B. (2020). The Southern origins of sustainable development goals: Ideas, actors, aspirations. *World Development, 126*, Article 104706.

Fulekar, M. H., Pathak, B., & Kale, R. K. (Eds.). (2014). *Environment and sustainable development* (p. 184). Springer India.

Gambe, T. R., Turok, I., & Visagie, J. (2023). The trajectories of urbanisation in Southern Africa: A comparative analysis. *Habitat International, 132*, Article 102747. https://doi.org/10.1016/j.habitatint.2023.102747

General Authority for Statistics (GAS). (2022). *General Authority for Statistics—Saudi Arabia.* https://www.stats.gov.sa/en

Grand, S., & Wolff, K. (2020). Assessing Saudi Vision 2030: A 2020 review. *Atlantic Council, 17.* https://www.atlanticcouncil.org/in-depth-research-reports/report/assessing-saudi-vision-2030-a-2020-review/

Hallegatte, S., Rentschler, J., & Rozenberg, J. (2020). Adaptation principles: A guide for designing strategies for climate change adaptation and resilience. *Climate Change Adaptation.* https://www.adaptation-undp.org/resources/reports/adaptation-principles-guid

Harlow, J., Golub, A., & Allenby, B. (2013). A review of utopian themes in sustainable development discourse. *Sustainable Development, 21*(4), 270–280.

He, L., & Zhang, X. (2022). The distribution effect of urbanization: Theoretical deduction and evidence from China. *Habitat International, 123*, Article 102544. https://doi.org/10.1016/j.habitatint.2022.102544

Heshmati, A., & Rashidghalam, M. (2020). Measurement and analysis of urban infrastructure and its effects on urbanization in China. *Journal of Infrastructure Systems, 26*(1), 04019030. https://doi.org/10.1061/(ASCE)IS.1943-555X.0000513

Hölscher, K., & Frantzeskaki, N. (2021). Perspectives on urban transformation research: Transformations in, of, and by cities. *Urban Transformations, 3*(1), 1–14. https://doi.org/10.1186/s42854-021-00019-z

Khakee, A. (2020). Planning dilemmas. *Planning Theory & Practice, 21*(1), 175–181. https://doi.org/10.1080/14649357.2019.1700074

Khan, M. B., & Iqbal, S. (2020). Vision 2030 and the national transformation program. In *Research, innovation and entrepreneurship in Saudi Arabia* (pp. 146–166). Routledge.

Kiamba, A. (2012). The sustainability of urban development in developing economies. *Consilience, 8*, 20–25.

Kisiała, W., & Rącka, I. (2021). Spatial and statistical analysis of urban poverty for sustainable city development. *Sustainability, 13*(2), 858. https://doi.org/10.3390/su13020858

Kramer, A. M., Emilio, R. I., & Cid, A. (2022). Resilience of urban systems: Perspective for the integration of climate resilience actions in the resilient cities network. In *The Elgar companion to urban infrastructure governance* (pp. 308–328). Edward Elgar Publishing.

Lange, J. (2017). *Population growth, the settlement process and economic progress: Adam Smith's theory of demo-economic development* (Doctoral dissertation, Paris 1).

Liu, H., Zhou, G., Wennersten, R., & Frostell, B. (2014). Analysis of sustainable urban development approaches in China. *Habitat International, 41*, 24–32.

Lu, Y., Song, W., & Lyu, Q. (2022). Assessing the effects of the new-type urbanization policy on Rural Settlement Evolution using a multi-agent model. *Habitat International, 127*, Article 102622. https://doi.org/10.1016/j.habitatint.2022.102622

Mandeli, K. (2019). Public space and the challenge of urban transformation in cities of emerging economies: Jeddah case study. *Cities, 95*, Article 102409. https://doi.org/10.1016/j.cities.2019.102409

McGranahan G. & Satterthwaite, D. (2014). *Urbanisation concepts and trends*. IIED Working Paper. IIED, London.

Ministry of Municipal, Rural Affairs and Housing. (2020). Gov.sa. Retrieved 14 July, 2022, from https://www.my.gov.sa/wps/portal/snp/agencies/agencyDetails/AC174/!ut/p/z0/04_Sj9CPykssy0xPLMnMz0vMAfIjo8zivQIsTAwdDQz9LQwNzQwCnS0tXPwMvYwNDAz0g1Pz9L30o_ArAppiVOTr7JuuH1WQWJKhm5mXlq8f4ehsaG6iX5DtHg4Ayo0duQ!!/

Nayebare, S. R., Aburizaiza, O. S., Siddique, A., Hussain, M. M., Zeb, J., Khatib, F., Carpenter, D. O., Blake, D. R., & Khwaja, H. A. (2022). Understanding the sources of ambient fine particulate matter (PM2. 5) in Jeddah, Saudi Arabia. *Atmosphere, 13*(5), 711. https://doi.org/10.3390/atmos13050711

Obeng-Odoom, F. (2010). "Abnormal" urbanization in Africa: A dissenting view. *African Geographical Review, 29*(2), 13–40.

Ochoa, J. J., Tan, Y., Qian, Q. K., Shen, L., & Moreno, E. L. (2018). Learning from best practices in sustainable urbanization. *Habitat International, 78*, 83–95. https://doi.org/10.1016/j.habitatint.2018.05.013

Oloke, O. C., Fayomi, O. S. I., Oluwatayo, A., Adagunodo, T. A., Akinwumi, I. I., & Amusan, L. M. (2021). The nexus of climate change, urban infrastructure and sustainable development in developing countries. *IOP Conference Series: Earth and Environmental Science, 665*(1), 012051. https://doi.org/10.1088/1755-1315/665/1/012051

Pilehvar, A. A. (2020). Urban unsustainability engineering in metropolises of Iran. *Iranian Journal of Science and Technology, Transactions of Civil Engineering, 44*(3), 775–785. https://doi.org/10.1007/s40996-019-00290-y

Pilehvar, A. A. (2021). Spatial-geographical analysis of urbanization in Iran. *Humanities and Social Sciences Communications, 8*(1), 1–12. https://doi.org/10.1057/s41599-021-00741-w

Poku-Boansi, M. (2021). Contextualizing urban growth, urbanisation and travel behaviour in Ghanaian cities. *Cities, 110*, Article 103083. https://doi.org/10.1016/j.cities.2020.103083

Potter, R. B., & Lloyd-Evans, S. (2014). *The city in the developing world*. Routledge.

Ramady, M. (2021). The Saudi capital market: Coming of age. In *Financial regulation and liberation* (pp. 43–57). Springer. https://doi.org/10.1007/978-3-030-68267-5

Rana, M., & Parves, M. (2011). Urbanization and sustainability: Challenges and strategies for sustainable urban development in Bangladesh. *Environment, Development and Sustainability, 13*(1), 237–256.

Rydin, Y. (2012). *Governing for sustainable urban development*. Routledge.

Saghir, J., & Santoro, J. (2018, April). Urbanization in Sub-Saharan Africa. In *Meeting challenges by bridging stakeholders*. Center for Strategic & International Studies.

Saudi Arabia Population Statistics 2022. (2022). *Global media insight*. Retrieved 17th July, 2022, from https://www.globalmediainsight.com/blog/saudi-arabia-population-statistics/

Sandoval, V., & Sarmiento, J. P. (2020). A neglected issue: Informal settlements, urban development, and disaster risk reduction in Latin America and the Caribbean. *Disaster Prevention and Management: An International Journal, 29*(5), 731–745. https://doi.org/10.1108/DPM-04-2020-0115

Serraglio, D. A., Ferreira, H. S., & Robinson, N. A. (2020). Climate-induced migration and resilient cities: A new urban agenda for sustainable development. *Sequência (Florianópolis)*, 10–46. https://doi.org/10.5007/2177-7055.2019v41n83p10

Shackleton, C. M., Cilliers, S. S., du Toit, M. J., & Davoren, E. (2021). The need for an urban ecology of the Global South. In *Urban Ecology in the Global South* (pp. 1–26). Springer.

Sileyew, K. J. (2019). *Research design and methodology* (pp. 1–12). IntechOpen

Spicer, J. (2021). Cooperative enterprise at scale: Comparative capitalisms and the political economy of ownership. *Socio-Economic Review*. https://doi.org/10.1093/ser/mwab010

Stiftel, B. (2021). Planners and the new urban agenda: Will we lead the agenda, or will the agenda lead us? *TPR: Town Planning Review, 92*(4). https://doi.org/10.3828/tpr.2021.13

Stoffel, B. (2017). *7 Countries with worse income inequality than the United States*. Retrieved 6 July, 2022, from https://www.fool.com/investing/2017/07/17/7-countries-with-worse-income-inequality-than-the.aspx

Tyburski, W. (2008). Origin and development of ecological philosophy and environmental ethics and their impact on the idea of sustainable development. *Sustainable Development, 16*(2), 100–108.

UN-Habitat. (2014). *The evolution of national urban policies—A global overview*. Unhabitat.org. https://unhabitat.org/the-evolution-of-national-urban-policies

UN-Habitat. (2015). *National urban policy: Framework for a rapid diagnostic*. Unhabitat.org. https://unhabitat.org/national-urban-policy-framework-for-a-rapid-diagnostic

UN-Habitat (2016a, October 20). *New Urban Agenda. United Nations Conference on Housing and Sustainable Urban Development (Habitat III) in Quito, Ecuador.* https://unhabitat.org/sites/default/files/2020/12/nua_handbook_14dec2020_2.pdf

UN-Habitat. (2016b). *Planning sustainable cities: Global report on human settlements 2009*. Routledge. https://unhabitat.org/sites/default/files/download-manager-files/Global%20Report%20on%20Human%20Settlements%202009%20Planning%20Sustainable%20Cities.pdf

UN-Habitat. (2018). *Tracking progress towards inclusive, safe, resilient and sustainable cities and human settlements*. http://uis.unesco.org/sites/default/files/documents/sdg11-synthesis-report-2018-en.pdf

UN-Habitat. (2019). *Saudi cities report 2019. Future Saudi cities program*. https://unhabitat.org/sites/default/files/2020/05/saudi_city_report.english.pdf

United Nations. (2015). Transforming our world: *The 2030 agenda for sustainable development (A/RES/70/1)*; General Assembly; United Nations. https://www.un.org/ga/search/view_doc.asp?symbol=A/RES/70/1&Lang=E

United Nations. (2018). *The sustainable development goals report* 2018. https://www.un.org/development/desa/publications/the-sustainable-development-goals-report-2018.html

Vourvachis, P., & Woodward, T. (2015). Content analysis in social and environmental reporting research: Trends and challenges. *Journal of Applied Accounting Research, 16*(2), 166–195. https://doi.org/10.1108/JAAR-04-2013-0027

World Air Quality Report. (2021). *IQAir*. https://www.iqair.com

World Bank. (2020). *Urban development*. https://www.worldbank.org/en/topic/urbandevelopment/overview

World Bank. (2022). *Cutting global carbon emissions: Where do cities stand?* World Bank Blogs. https://blogs.worldbank.org/sustainablecities/cutting-global-carbon-emissions-where-do-cities-stand

World Cities Report. (2022). *Envisaging the future of cities*. UN-Habitat. https://unhabitat.org/wcr/

Ye, X., & Niyogi, D. (2022). Resilience of human settlements to climate change needs the convergence of urban planning and urban climate science. *Computational Urban Science, 2*(1). https://doi.org/10.1007/s43762-022-00035-0

Yigitcanlar, T., & Teriman, S. (2015). Rethinking sustainable urban development: Towards an integrated planning and development process. *International Journal of Environmental Science and Technology, 12*(1), 341–352.

Zeng, X., Yu, Y., Yang, S., Lv, Y., & Sarker, M. N. I. (2022). Urban resilience for urban sustainability: Concepts, dimensions, and perspectives. *Sustainability, 14*(5), 2481. https://doi.org/10.3390/su14052481

Chapter 3
Sustainable Development Fantasies and Realities in Saudi Arabian Cities: Urbanization-Environment Analysis

3.1 Introduction

The surge of interest in Global South urbanization discourse since the turn of the twenty-first century demonstrates urban studies and planning's increasing awareness that urbanization is the prime mediator between urban sustainability and urban growth induced environmental challenges (Mutuku, 2018; UN-Habitat, 2019, 2020, 2022; World Economic Forum, 2022). Urban planners and scholars across many Global South cities have led efforts to link and demonstrate the social, physical, economic, and environmental systems that shape and are framed by urbanization. They encourage sustainable development approach that analyzes how urbanization patterns produce and compound inequity and uneven forms of urban growth and development by examining understandings of urbanization origins, occurrence, and impacts (UN-Habitat, 2019, 2020). Such an approach allows for the engagement of the metabolic character of urbanization, as it reveals urbanization's role in the production of dysfunctional and distorted urban spaces (World Economic Forum, 2022).

The term "urbanization" was initially understood as a universal concept of large-scale clustering of human populations inside multifunctional settlements (Fang & Yu, 2017). For example, the World Economic Forum (2022) characterized urbanization in the context of population growth, indicating that over 55% of the global population, accounting for 4.3 billion people is concentrated in regions defined as urban, and the figure is expected to increase to nearly two-thirds by 2030. The "rapid urbanization" term has now evolved to include the dynamics of spatial extent of cities, and proliferated widely through environmental, climate change, geography, urban studies, and planning research across the global south (e.g., Abubakar & Dano, 2020; Cobbinah et al., 2015; UN-Habitat, 2019). Dawson et al. (2017) describe the development of urbanization as being among the most prominent and apparent examples of anthropogenic activity on the planet. Such rapid urban growth and urbanization trends have several concerning consequences for society, the economy, and the environment.

A. I. Almulhim and P. B. Cobbinah, *Urbanization Rethought*, Urban Sustainability,
https://doi.org/10.1007/978-981-92-0118-1_3

Abubakar and Dano (2020) and Almulhim and Abubakar (2021) identify various urbanization-induced environmental problems in Saudi Arabia's major urban development categories such as ecology, habitat loss, poaching, growing desertification, and soil erosion, which have contributed to pollution, climate change, deforestation, and slum expansion.

Contemporary sustainable development in Global South cities attempts to accomplish many things: drawing on the sustainable development goals (SDGs), a global agenda that champions a multidimensional approach to improving sustainability, equity, and resilience issues. The SDGs, particularly Goal 11 and the New Urban Agenda conjure a series of urban response measures that improves equity, inclusiveness, security, resilience, and long-lasting cities. These objectives clearly represent the need to address both current and future urban pathologies, and to prioritize urbanization as part of an international development policy (UN-Habitat, 2019). Sustainable development in Global South cities is broadly concerned with managing rapid urbanization and addressing its implications thereof. It foregrounds rapid urbanization challenges such as urban sprawl, ecological disturbance, land-use distortions, the messiness of unregulated informality, housing unaffordability, energy inadequacies, infrastructure deficits, and inadequate essential services like education, healthcare, and transportation, which will be exacerbated when the predicted trends in future urbanization occur. Urban poverty, unemployment, violence, and pollution are only some of the concomitant issues that need to be addressed (Elimam, 2022; UN-Habitat, 2020).

In other words, sustainable development in Global South cities considers cities economic and social growth potentials (Colenbrander, 2016), production efficiency (Shaban et al., 2022; The World Bank, 2020), and environmental integrity (Cobbinah et al., 2015). As global population doubles, the planet's urban population triples and rapid urbanization becomes a more common feature of Global South Cities, a new concern of urban environmental quality emerges (World Health Organization [WHO], 2021). This urban environmental concern must be assessed toward sustainable development agendas, and the urgency of addressing the socio-spatial reconfiguration of rapid urbanization geopolitical landscapes (Chen et al., 2014). As several writers have demonstrated, rapid population growth and urbanization phenomena are paraded as socio-economic development stimulus, and cities worldwide significantly contribute to the global economy as generators of new ideas and products for both the international economy and outlying areas of their respective nations (Chen et al., 2014; The World Bank, 2020). However, they are not free from sustainability tensions, as they also create and exacerbate pre-existing environmental sustainability tensions, including environmental pollution, destruction of ecologically sensitive areas, and loss of urban green spaces.

Despite a solid foundation built on the normative principles of sustainable development, academic formulations and policy prescriptions have still accorded limited attention to environmental quality in Global South cities. As such, they fail to consider how ontological and epistemological environmental quality issues are embedded in urbanization patterns and management. The world is still urbanizing, especially in

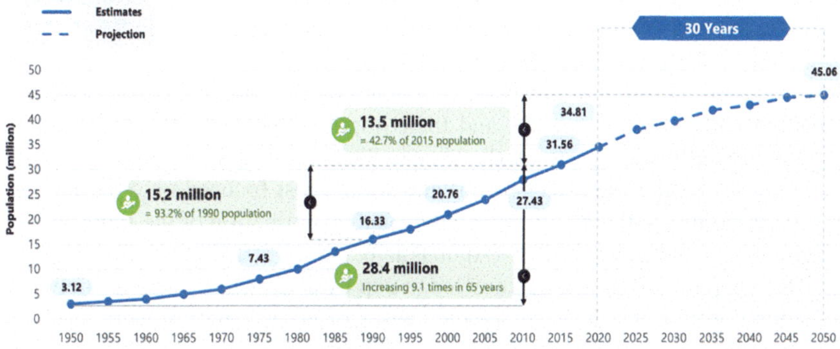

Fig. 3.1 Saudi Arabia's population increase between 1950 and 2050. Authors' computation using data extracted from UN-Habitat (2019) and General Authority for Statistics (2022)

Global South countries, and the urgency to understand and address urban environmental quality issues is growing. For example, Saudi Arabia has more than 80% of its population residing in cities (Almulhim et al., 2022). Saudi Arabia's population has more than doubled over the past 25 years and has grown by 2.5% annually since (see Fig. 3.1). At 46.2 million, the Saudi population is projected to reach its peak in 2065 and decline to 44 million by 2100 (UN-Habitat, 2019). Saudi Arabia's fast economic growth, largely due to its oil reserves, has enabled huge investments in infrastructure, education, and health; this phenomenon is thus a key determinant of the country's population expansion. Additionally, the boom in investments has contributed to an enormous influx of foreign migrants. Saudi Arabia hosted 13.12 million overseas migrants in 2019, a 12.8% increase between 1995 and 2019 (Migrant Refugees, 2020).

While fast urban growth, infrastructure building, and societal transition in Saudi Arabia have resulted in a dramatic rise in living standards in urban centers (Chen et al., 2014), they have also created significant urban environmental problems. In response, the Saudi government has launched several sustainable urban development programs related to the following: urban greenery, tourism promotion, youth employment, economic development, alternative energy, education on economic cities, a public transit plan, and pollution investigation and management (Aina, 2014; Alqahtany et al., 2014). In 2013, the government execute the Future of Saudi Cities initiative by way of the Ministry of Municipal and Rural Affairs and in partnership with UN-Habitat. Sustainable urban growth in Saudi Arabia is a goal of the Future of Saudi Cities Program (FSCP), which aims to achieve this goal by improving the institutional framework and urban legislation, as well as by strengthening the management and planning of economically efficient cities (UN-Habitat, 2019). Similarly, the Saudi Green Initiative was launched as part of Saudi Arabia's Vision 2030, and its first wave of over 60 initiatives represents an investment of over SAR 700 billion toward the development of a green economy. The initiative's three overarching targets are to reduce carbon emissions by more than 278 MTPA by 2030, plant 10 billion trees, and increase the percentage of marine and terrestrial protected areas in Saudi Arabia to more than 30% (Saudi Green Initiative, 2022).

However, limited research has been conducted on the foundational ties between urbanization and the environment, and their implications for urban sustainability across Global South cities, particularly those in Saudi Arabia. While multiple studies have examined broadly the effects of urbanization, most focused on environmental and socio-economic concerns, limiting a thorough examination of the links between urbanization and the environment (Abubakar & Aina, 2018; Abubakar & Dano, 2020). This chapter addresses the knowledge gap by analyzing the environment and urbanization correlations. It discusses the causes of urban environmental problems and the contribution of anthropogenic factors. It argues that unless rapid urbanization and its environmental implications are countered by sustained economic expansion and environmentally responsible building practices, sustainable growth will be an illusory mirage for Global South countries like Saudi Arabia.

3.2 Environment-Urbanization Nexus: Toward Sustainability

The concept of "the environment" can be understood according to multiple perspectives and definitions. For example, according to Byjus (2020) the environment can be defined as the whole of all factors, both living and non-living, that impact human life. On the contrary, Alturki (2015) considered the environment a phenomenal non-specific concept encompassing both human and non-human variables, and that, noteworthily, every living and non-living creature in a specific environment forms part of a diverse network that shapes their shared environment. Nasrudin (2022) on the contrary, defined the environment as a dynamic entity that varies between community groups; conditions comprise not only physical settings but also social and cultural aspects, such as population size, culture, traditions, lifestyle, tastes, and customs.

Urban sustainability enhances human standards of living, safeguards the natural ecology, and protects natural resources for future generations. Based on this discussion, sustainability refers to the environment's capacity to satisfy the fundamental needs of living and non-living components of environmental, sociocultural, and economic structures. It is essential that the current and future needs of these elements are recognized and planned for. In other words, sustainability refers to the environment's ability to be self-sufficient (Evans, 2020). Jianping et al. (2014) presented evidence that urban development, homes, land use, infrastructure development, and planning are merely a few fields that have prioritized elevated environmental sustainability in response to the threat posed to the planet's carrying capacity by the current state of environmental degradation (Bibri et al., 2020; Salama & Alshuwaikhat, 2006; The World Bank, 2018). Nassar and Elsayed (2018) reported that sustainability must not be viewed as a purpose for property or urban agendas because ineffective housing or urban programs are considered sustainable while disaster risks, along with other social, economic, and environmental challenges, are ignored. In fact, sustainability can be impeded by sustainable housing or urban programs. Recent evidence, on the contrary, implies that sustainable urban growth and housing programs are intricately intertwined with environmental protection (Aliu, 2022; Wan, 2013).

The objective of environmental protection is thus to sustain a natural environment that is in good health and to restore it when necessary. Sustainable development, by contrast, is concerned with social, environmental, and economic goals to achieve long-term growth that is equitable and beneficial to both the present generation and those to come (National Audit Office, 2015). According to the United Nations (2021) sustainable development entails the capability of urban areas and the environment to continue operating per the intended standard of living by communities and starts with national policies and regional development plans. This is accomplished without limiting available alternatives to current and future generations, and without precipitating any detrimental effects. However, to achieve this goal, policymakers must comprehend the expanding ecological and social "footprint" of urbanization and declining number of untouched regions left on Earth (Galli et al., 2020). The sizable footprint of metropolitan areas on the environment, which is not exclusive to emerging nations, may be accounted for by shifts in the ecological systems of urban environments caused by expanding populations, overcrowding in human settlements, and unregulated natural resource exploitation.

The environmental issues listed in Fig. 3.2 exert destructive socio-economic and environmental effects (UN-Habitat, 2019). The negative psychological attitude of urban residents toward the environment; a failure in understanding and a lack of awareness concerning climate change, urban sustainability and its causes and impacts on humanity (Almulhim, 2021, 2022a, 2022b); and ineffective management techniques that impact the environment (Barbora, 2019) are all factors that contribute to the problem. Crucially, however, urbanization undoubtedly benefits the economy as it develops industrialization, provides social benefits, improves rural development, and caters to technological advancement. However, urbanization is not without major cons in terms of the environment, which can directly impact the economy, public health, and living standards. Most environmental issues are caused by the processes involved in development, and their consequences are evident on the global, national, and local scales. These effects are the outcomes of human activity and almost always negatively impact people, their means of subsistence, and animal and plant lives, either at present or in the future (Abubakar & Aina, 2018; Abubakar & Dano, 2020). This has repercussions that will be experienced for a prolonged period in the sustainable development context, particularly due to worsening economic prospects. Thus, concerns for the urban environment will predominate on the agenda for sustainable development in emerging countries for the foreseeable future. Further steps must be taken, existing policies must be supported, and awareness regarding the implications of urbanization for sustainability and the environment must be recognized. This chapter draws insights from the environment-urbanization nexus and their implications on urban sustainability to understand how urbanization in Saudi Arabia is dictating, framing, and producing urban environmental quality concerns.

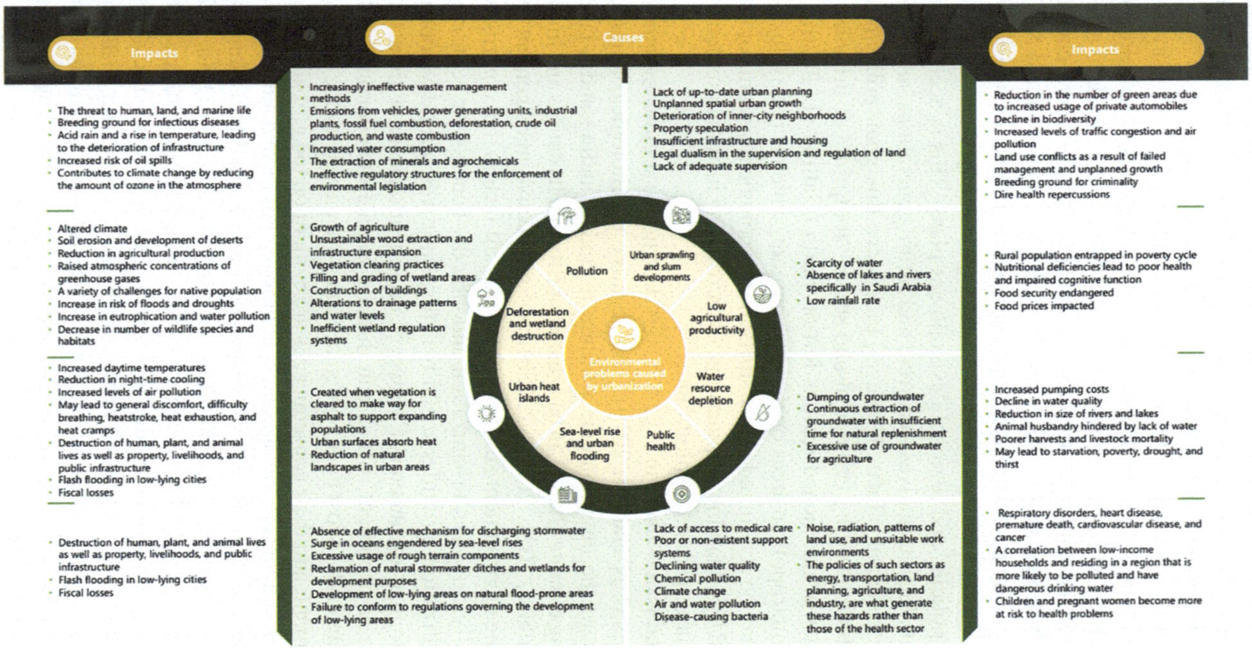

Fig. 3.2 The causes and impacts of environmental problems. Illustration created by the authors; *Sources* United States Geological Survey (2003), Tolba and Saab (2009), UN-Habitat (2011), Liu et al. (2014), Revi et al. (2014), DeNicola et al. (2015), Belloumi and Alshehry (2016), Zhang (2016), Abubakar and Aina (2018), Mahmood et al. (2019), UN-Habitat (2019), Abubakar and Dano (2020), Mahmood et al. (2020), Davies (2021), Almulhim and Abubakar (2021), Almulhim (2021), Koroso et al. (2021), US EPA (2021), and Almulhim and Aina (2022)

3.3 Research Approach

3.3.1 Research Context

This chapter focuses on Saudi Arabia. In 2022, the population of Saudi Arabia was estimated to be 34.11 million, with 84.7% or 30.36 million people living in urban areas (General Authority for Statistics, 2022; Worldometer, 2022). Most urban migrants opt to reside in Saudi Arabia's major cities, such as Riyadh, Dammam, Mecca, and Jeddah. The size of the country's total population is expected to expand and reach its peak of 45.35 million by 2060 (World Population Review, 2022). The nation, which is administratively split among 13 regions, has a population density of 16.67 people per square meter as of 2022 and a median age of 32.4 years. On the other hand, the country's gross domestic product (GDP) is 700.1 billion US dollars in 2020, representing 0.62% of the global economy, placing it as the fourth most prosperous Arab country and 15th in the world (The World Bank, 2022a). Saudi Arabia was selected for this research because of (i) its rapid urbanization levels; (ii) evidence of increasing concentration on socio-economic variables with limited focus on environmental implications; (iii) evidence of increasing concerns of environmental quality; and (iv) availability of basic urbanization and sustainable development data.

3.3.2 Data Collection and Analysis

This chapter involves extensive literature review on urbanization, environment, and sustainable development. The examination of secondary data, as well as a desk-based review of the pertinent literature that addresses the research objective, constituted the data gathering process (ACAPS, 2014). The first step was to classify and retrieve pertinent literature using Google Scholar and Crossref. The search keywords included "urban sprawl," "sustainable consumption," "city management," "infrastructure development," "governance," "allocation of employment options for burgeoning urbanization," "environmental pollution," "climate change adaptation," and "insufficient urban services" among other similar topics. The researchers collected data from websites as well as online resources such as journal articles, technical reports, books, web content, and conference reports.

A content analysis of the qualitative data was used to perform the data analysis, and the results were linked to anthropogenic environmental problems due to the production of goods and services in Saudi Arabia's main cities (Shava et al., 2021). The content analysis method utilizes a number of iterations that occur over two steps. First, all materials were sorted into categories based on the above search criteria, even though some fit into more than one category. Subsequently, each resource underwent a methodical search to verify the themes that emerged or to find fresh viewpoints until there was nothing more to look for. In the end, a synthesis of the themes was performed by locating their interrelationships, as well as the points at which they overlapped or differed.

3.4 Urbanization-Induced Environmental Quality Threats in Saudi Arabia's Cities

3.4.1 Understanding Pollution (Water, Air, and Land) in Urbanizing Country

Given the desalination process that Saudi Arabia uses on its water supply, the water not only has a high concentration of salt, but also a significant concentration of metal (Sharkh et al., 2022). The water is unfit for use in either the agricultural sector or consumption in cities, much of it is simply discharged back into the ocean, negatively impacting on the marine ecosystem. Water pollution is the cause of the demise of hundreds of marine organisms and exerts a negative impact on the biological diversity of some of Saudi Arabia's seas, such as the Gulf of Aden, the Red Sea, and the Arabian Sea (Im, 2012).

The burning of crude oil is the primary source of air pollution in Saudi Arabia since its economy heavily relies on the export and import of oil (see Fig. 3.3; EIA, 2021). During the process of refining crude oil, various types of greenhouse gases are produced (United States Energy Information Administration, 2000) that are partly responsible for the rise in temperature observed in Saudi Arabia. Desertification is an additional consequence of air pollution (Freudenrich, 2022; gov.sa, 2021). Saudi Arabia has some of the most polluted air globally, with 108 $\mu g/m^3$ of PM2.5 which is above the limits of the tolerable standard of 10 $\mu g/m^3$ of PM2.5 set by the WHO (2021).

Waste dumps and oil leaks are two of the main contributors to Saudi Arabia's land pollution problem, which is exacerbated by the country's rapid industrialization and urbanization, and rising population (Zafar, 2021). Approximately 53 million tons of garbage are generated in Saudi Arabia each year, poisoning the country's soil and groundwater (International Trade Administration, 2022). The National Center for Waste Management in Saudi Arabia reported a $1.3 billion price tag for environmental resulting from solid waste in the year 2021. A little over a third of the country's garbage is generated in only three cities: 21% from Riyadh, 14% from Jeddah, and 8% from Dammam (International Trade Administration, 2022). Untreated garbage from cities is mostly dumped in landfills, while some is burned.

The public transit system in Saudi Arabia is not nearly as developed as it should be. As a result, a significant factor in elevated air pollution is the use of private cars. This trend contributes to dangerously high levels of heavy metals present in the soil of urban areas. These metals are toxic to humans as well as plants because they cause soil contamination, which stunts plant growth (Kadi, 2009). Unfortunately, pollution

Fig. 3.3 Pollution caused by the oil and gas industry. *Source* Al Shammary (2020)

is a major threat to achieving the SDGs, particularly Goal 3 on good health and well-being, Goal 6 on clean water, and Goal 13 on climate action.

3.4.2 Urbanization and Water Resource Depletion

Rapid urban growth has detrimental impacts on air quality, energy use, land quality, and the water quality of river basins through alterations to their topology, hydrology, and geology (see Fig. 3.3; Wilson & Chakraborty, 2013). Saudi Arabia's water scarcity is the outcome of a significant drop in precipitation, high rates of evaporation, and excessive pumping from subterranean water tables (DeNicola et al., 2015). Considering that freshwater sources are undergoing a general decline of 241–1435 million cubic meters per year, Saudi Arabia is among the most water-stressed regions in the world, particularly when compared to a recharge amount of about 3850 million cubic meters per year. (Odhiambo, 2017). This is primarily attributable to rising temperatures as well as rising demands for irrigation, which is anticipated to rise by 5%–15% by the year 2050 (Chowdhury & Al-Zahrani, 2013). The fact that there is an approximately 6% growth in residential demand for piped water for the past four decades further exacerbates the problem (Abubakar & Aina, 2018). Twelve years ago, the nationwide average for water usage was 235 L per person per day, but more than 300 L per person per day in metropolitan areas. This amount of water use is equivalent to that of urban hubs in developed nations that have an abundance of freshwater ecosystems (Gazzeh & Abubakar, 2018). This is in part due to urbanization as well as other factors like wasteful attitudes, wealthy lifestyles, and substantially subsidized water costs (Abubakar & Aina, 2018). In this case the responsible consumption emphasized by SDG 12 is under threat.

3.4.3 Urbanization's Implications on Urban Heat Islands

The term "urban heat island" (UHI) refers to a localized region in a city where temperatures are often significantly higher compared to surrounding rural areas. It is a condition of extreme surface temperatures that occur when built-up urban zones start generating and preserving heat due to massive building mass, impenetrable built surfaces, and heat generated because of human activities, in addition to the inadequacy of water bodies and vegetation for the absorption of heat (Abubakar & Bununu, 2020). Human thermal stress, fluctuations in relative humidity, and higher energy bills are merely some of the issues caused by this phenomenon (Santamouris, 2013). The UHI impact results from the alteration of the urban boundary layer's atmosphere and thermophysics as a direct outcome of human activity on surrounding natural landscapes. UHIs are caused by a combination of factors, such as population density in urban regions and vegetation fractions, and mainly urbanization (Mohajerani et al., 2017). Air temperature has risen by almost 4 °C in the past few decades, with a projected temperature rise along the coastal areas from 2 to 2.75 °C. Similarly, the summertime temperature in some urban areas can reach 52 °C (Abubakar & Aina, 2018). This is however expected as CO_2 emission per capita in Saudi Arabia alone has been increasing from10.25 mt in 1990, 11.82 mt in 2000, to 15.27 mt in 2018; indicating more than half increment in the average for high-income nations (11 mt/capita) (The World Bank, 2022b). It is possible that Saudi Arabia will warm more than the global average by 2040, with projections ranging from 2.2 to 2.7 °C (Darfaoui & Al Assiri, 2010). Extreme weather, such as heat waves and sandstorms, will become more common in metropolitan areas due to climate change and limited green space cover in these places (see Fig. 3.4).

3.4.4 Sea-Level Rise and Urban Flooding

The primary contributors to sea-level rise are two phenomena directly linked to anthropogenic global warming: water released from ice sheets and glaciers as they melt, followed by the expansion of seawater as it heats (NASA, 2022). Ninety-four percent of the buildings at risk from coastal floods result from urbanization (Mason et al., 2020). Sea levels are anticipated to rise due to climate change which will have repercussions for low-lying coastal community buildings and other forms of infrastructure. Saudi Arabia has a 2640-km coastline and according to Tolba and Saab (2009) 30.3% of the population lived within 100 km of the shoreline in the year 2000. Hence, the coastal cities of Dammam and Jeddah are currently facing multiple instances of flooding. For instance, in 2009, the city of Jeddah was hit by the greatest flood it had seen in the preceding 30 years. The storm resulted in 122 fatalities, 37 people going missing, and 3861 dwellings being inundated and destroyed completely (Al-Maamary et al., 2017). The failure to develop adequate stormwater drainage systems to collect runoff is a factor that further contributes to

Fig. 3.4 Areas at risk of the UHI effect and sandstorms, Dammam. *Source* UN-Habitat (2019)

the dangers posed to the nation's food supply (Aina, 2017), and further threatens the attainment of SDG 11.

3.4.5 Low Urban Agricultural Productivity

Agricultural land is continuously being lost due to urbanization. This happens directly when land is acquired for urban use, and indirectly when agricultural land is used for unproductive rural activities such as leisure, leisure farming, or horse rearing (Beckers et al., 2020). Farmers are put under stress as a result of urbanization processes, which also make farming activities more difficult because of less agricultural land, increased competition for land, and negative externalities. Millions of people worldwide are at greater risk of becoming homeless, facing starvation, and food insecurity because of the consequences of climate change (UN-Habitat, 2011). Due to climate change, harvests of wheat, vegetables, fish, and dates have decreased in Saudi Arabia, which has led to a rise in the cost of food in local markets. Several different plant varieties are unable to survive in most habitats due to current high temperatures, which are expected to persist in the future. In addition, approximately a third of vegetable production happens in greenhouses; the other proportions are impacted by temperature swings and inadequate rainfall. Tolba and Saab (2009) report that approximately 80 animal species are in danger, and it is anticipated that the annual rainfall will decrease by 20%–25% in the region. In a similar vein, aquatic creatures are increasingly confronted with a reduction in the quantities of oxygen in their environments, with a concurrent rise in the amount, probability, and magnitude of acidification.

3.4.6 Urban Sprawl and Slum Development

SDG 11 focuses on creation of resilient, sustainable, and inclusive cities. Yet, some of the most significant concerns that have arisen because of rapid urbanization and growth of population include urban sprawl and low-density development in slums. The role of central residential districts is somewhat diminishing due to leapfrog development and the expansion of scattered communities on the outskirts of major urban hubs in Saudi Arabia. Saudi Arabia currently has an average population of 42 people per square mile, or 16 people per square kilometer (Worldometer, 2022). Based on the findings of Aina et al. (2008) Riyadh has expanded at a pace unparalleled by any other city, growing from a traditional town that was once only a single square kilometer in size to a metropolis that is three thousand times that size in a little over a century. Dammam stands out because its physical size increased by more than four times between 2004 and 2017, yet its population grew by less than half during the same time. Riyadh's population increased by 2.5 times between 1990 and 2014, while the city's physical footprint expanded by more than three times. Between 1990 and 2014, Riyadh's built-up area increased by 95.861 hectares from 30,305 hectares, or 9.4% each year (UN-Habitat, 2019). The city's built-up area grew by around 5% between 1990 and 2000, and by 8% between 2000 and 2013, indicating that Riyadh has grown at a rapid rate (see Fig. 3.5). Even medium- and small-sized cities in Saudi Arabia are experiencing urban sprawl. Ha'il, for example, with a population of less than 400,000, grew by more than 3.5 times between 2004 and 2019 (UN-Habitat, 2019). Owing to their expansive natures, Saudi cities exert a detrimental effect on the environmental footprint, which can be defined as the number of resources required by each individual. The ecological footprint that humans have left behind in Saudi Arabia is presently three times larger than what the environment can sustain (Aina, 2014).

As presented in Fig. 3.5, growth borders have been expanded as a tool for urban growth management, particularly in big metropolitan regions. Other measures for managing urban growth, including suitable densification regulations, satellite towns, and new cities, have been overlooked as a result. As mentioned earlier, 66% of the urban growth borders were expanded between 2000 and 2013 in Riyadh's built-up area, while densification or infills accounted for 24%. A large, continuous urban mass can be found in Riyadh, Madinah, and Jeddah; long urban fingers can be found in Makkah; and in Dammam, the urban pattern is highly fractured and sprawling due to the expansion of the urban boundary (see Fig. 3.6). The rapid growth of these

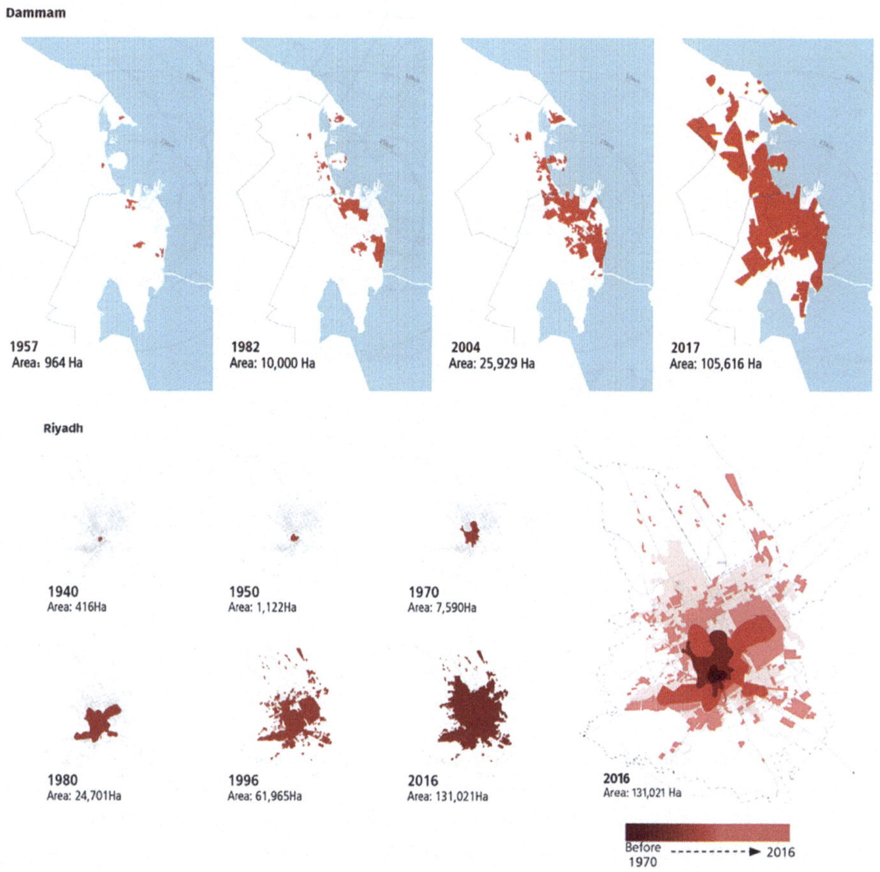

Fig. 3.5 Urban growth stages for Dammam and Riyadh. *Source* UN-Habitat (2019)

cities and resulting urban sprawl have created significant new issues for municipal planners tasked with managing urban growth. In addition, residents living on the edges of cities face significant travel times to reach the city center, causing increases in traffic and service delivery issues (UN-Habitat, 2019).

3.4.7 Deforestation and Wetland Destruction

Deforestation is caused in part by urbanization, the expansion of agricultural land, and the consumption of energy. Wood is an essential natural resource that people near forests utilize. People living in traditional dwellings in rural areas use wood inefficiently on a regular basis (Al-Subaiee, 2016), and food insecurity is expected to worsen. In response, there has been an increase in the rate at which forests are cut

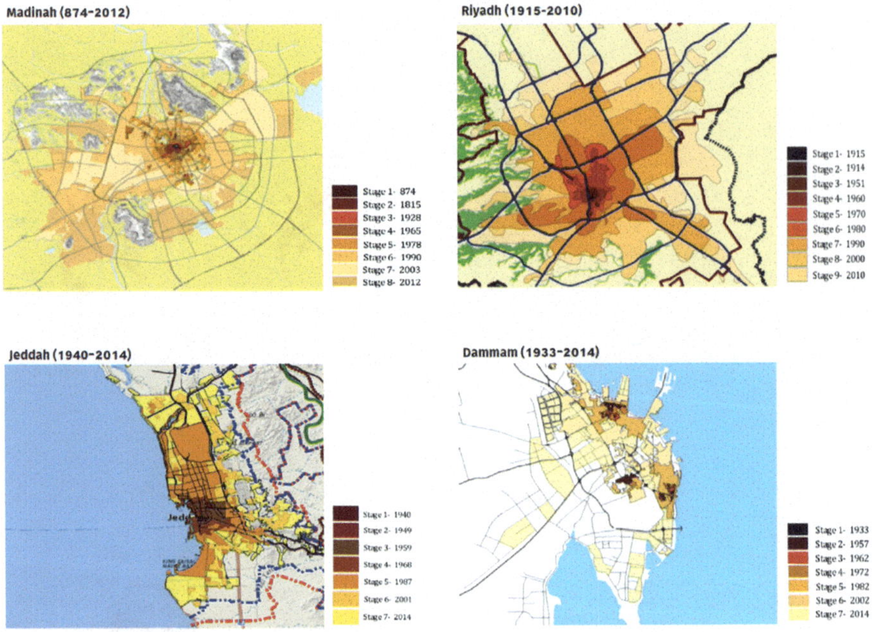

Fig. 3.6 Patterns of urban growth in Saudi cities. *Source* UN-Habitat (2019)

down to create room for more agricultural land. Both legally and illegally felled trees, however, contribute to deforestation. According to studies on the preservation of woodlands, there are very few efforts currently underway to reduce the rate of deforestation (Al-Subaiee, 2016). Likewise, as a type of ecosystem widely acknowledged for its significance in sustainable development and humanity's well-being because of the social, economic, and environmental importance that they possess, wetlands are very susceptible to the effects of global environmental change (IPCC, 2007); it will have a considerable impact on the ecosystem services provided by wetlands, namely water quality and supply, wildlife habitats, fisheries, and recreational activities. The long-term persistence of Saudi Arabia's wetlands is in jeopardy due to human activity causing continued habitat degradation, population growth, and coastal development. Amid the ecological functions and societal benefits that they provide, Saudi Arabia's wetlands face substantial problems because of these activities (Al-Obaid et al., 2017; see Fig. 3.7).

3.4.8 *Urbanization as a Threat to Public Health*

The adverse impacts of climate change on the well-being and health of civilizations are primarily because it exacerbates pre-existing health problems (Revi et al., 2014). Urbanization has been linked to a number of health concerns, comprising, but

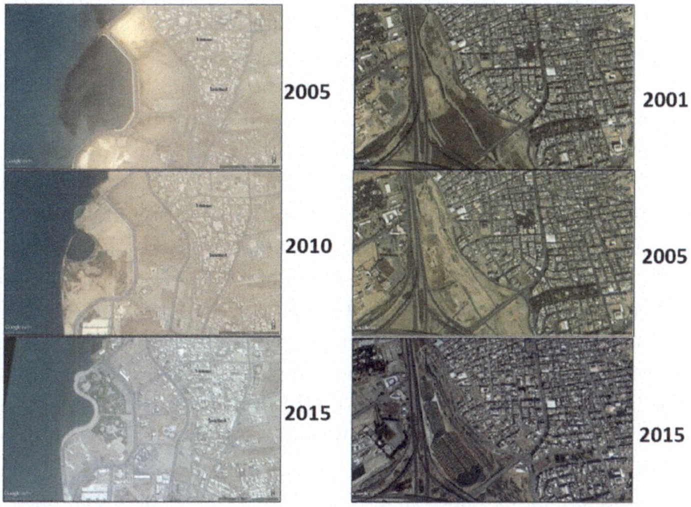

Fig. 3.7 Overview of wetlands in Saudi Arabia. *Source* Al-Obaid et al. (2017)

not restricted to, malnutrition, respiration, infectious diseases, and unsanitary living conditions (Kuddus et al., 2020). The nutritional health of low-income populations is severely compromised by urbanization. Low-income residents in urban areas often eat poorly balanced diets that leave them susceptible to illness and poor nutrient absorption, as well as a concomitant loss of appetite (Kennedy, 2009).

Epidemics caused by air pollution, as well as exposure to ultraviolet and infrared light, are some of the climate change-related concerns present in Saudi Arabia. There have been reports of particular allergens in the air becoming more concentrated at regular intervals, which can lead to respiratory illnesses and allergic reactions (Dano & Alqahtany, 2019). According to Tolba and Saab (2009), the number of disability-adjusted life years for every 100,000 people in the country is the estimated outcome of climate change regarding the burden of disease in Saudi Arabia. Moreover, based on the work of Brauer et al. (2016) a Saudi person's life expectancy is reduced by around one-and-a-half years due to exposure to air pollution.

3.5 Discussion: Realities of Sustainable Development in Saudi Urbanization

The concept of sustainable development has been gradually adopted by several key initiatives across Global South countries, including Saudi Arabia, such as the Transformation Program 2020 and Vision 2030, in response to rapid urbanization and burgeoning severity of environmental issues. This trend strongly emphasizes the significance of preserving and enhancing the national living standard by guaranteeing that each decision made in this direction incorporates social, environmental, and economic ramifications. Progress—despite varying—has been made in the application of sustainable development in Global South. Yet, this research shows that considerable commitment is required to achieve urban sustainability. By using urban sustainability model analysis and based on cases in Saudi Arabian cities, this research indicates that rapid urbanization is framed as conflictual problematic by actors who consider its occurrence as constituting a threat to urban environmental quality on three different levels. One way in which this occurs is when the process surrounding urbanization is perceived as sustainably flawed (over concentration on socio-economic issues). A second context in which urbanization is perceived as problematic and conflictual is when it is regarded as posing a threat to urban environmental values (disregard for environmental resources). And third, a sense of conflict also emerges from a perception of urbanization-incompatibility with the established land-use order (subsuming of conventional land uses due to rapid urbanization).

In Saudi Arabia, the eight issues discussed show considerable evidence to significant urbanization conflict on environmental values, and the conflict on issues was experience across multiple cities in the country. Environmental quality conflicts and problems appear to have emerged primarily as a result of urbanization process and the establishment socio-economic agenda and was exacerbated by inadequate framing of urban sustainability as it is overly socio-economically aggressive and exerting undue pressure and existential influence on urban environmental resources. It is interesting to note, however, that sustainable development policies in Saudi Arabia failed to frame the urbanization process in environmental bureaucratic and destructive terms, thus disregarding the centrality and criticality of urban environmental aspects of rapid urbanization. Disputes on urbanization emerged not around socio-economic prospects but primarily around questions of urban sustainability of important environmental resources such as water, agricultural land, and ecosystems.

Environment theory (Byjus, 2020) offers one explanation of the socio-economic impact of the urbanization in Saudi Arabian cities. All the major cities experiencing rapid urbanization appear to have been evaluating sustainable development in terms of economic costs and benefits they expected to sustain the city. As a result of the varied environmental management concerns, all major cities considered urbanization's socio-economic benefits to outweigh its environmental cost and therefore increased investment in socio-economic agendas. The challenges associated with urbanization, developmental issues, consumption patterns, urban production, and urban residents' psychological perceptions are merely some of the factors contributing to the complex nature of Saudi Arabia's urban environmental problems. By all accounts, environmental issues have received scant attention from Saudi Arabia's numerous government administrations.

With the general wave of global interest and commitment toward sustainable development, especially in the Global South, the current Saudi government, focuses on environmental concerns, and various programs have been formed to safeguard the planet's natural ecosystems and dwindling supplies of raw materials. However, these programs and initiatives, just like in other Global South cities, are in their initial stages; hence, they require improved research and implementation efforts to ensure these programs' effectiveness, as well as environmental protection and sustainability. Guidelines for the implementation of sustainable urbanization policies to conserve the environment and its resources have become urgent to guarantee the success of such initiatives on the part of the government and policymakers. Given the growing number of citizens settling in metropolitan areas, there is a lack of available housing and infrastructure to accommodate them. The increased population density in metropolitan regions of Global South countries has increased the amount of pressure placed on the available land for crop availability, infrastructure development, home construction, and the manufacturing of goods, especially in Saudi Arabian cities. Therefore, an ecological disparity has been created, accompanied by negative environmental outcomes, including risks and catastrophes (Baus, 2017).

Most Global South cities are witnessing major building and infrastructure development projects. Such expansions must be adequately planned to make sure that there are no conflicts between humans and wildlife, and that natural habitats are maintained to the greatest practicable extent. Unfortunately, across Global South countries including Saudi Arabia, rapid urbanization has contributed to the loss of biodiversity and harm to local animals, disrupting their native habitats and even affecting their conservation status (Alatawi, 2022). The systematic extinction of diverse animal species necessary for the formation and maintenance of a wide range of ecological cycles and food webs is leading to an ecological imbalance, even posing a threat to the continued existence of humans in their natural habitats (Weiskopf et al., 2020).

The social, economic, and environmental welfare of Global South urban populations will become increasingly intertwined in the years ahead. For urban and regional policymakers and planners to fully comprehend the forces underlying the systems at play in urbanization, it is crucial that accurate, ongoing, and prompt information be disseminated on national patterns in urban development and the growth of cities in time and space (Li & Zhou, 2017). Since Saudi Arabia's economy and society rely heavily on the country's ecology for power generation, food production, and the availability of raw materials for industrial activity, the deterioration of the environment is likely to worsen unless corrective measures are taken.

Within the foregoing context, how can Global South cities effectively improve urban environmental sustainability in the face of rapid urbanization in terms of policy. This chapter argues for a policy shift away from highly concentrated urban areas toward more evenly distributed populations throughout the country. This would expand the development scope, particularly in tourism and mining, thereby maintaining economic vitality in underdeveloped rural and frontier regions with great population growth potential. Planned city infills and implanting the densification policy in cities is paramount, as this enables the efficient exploitation of current services and infrastructure and strengthens the population-housing capacity of small and medium-sized settlements.

3.6 Conclusion

This chapter intended to explore the urban environmental conundrum associated with rapid urbanization with the context of sustainable development. Eight key knowledge areas of urbanization-environment implications were discussed. Findings show the significance of the connection between the environment and urbanization. There is a need for proper knowledge of this correlation to establish programs and strategies that will efficiently and effectively stimulate sustainable environmental management. The process of urbanization in Saudi Arabia, and many Global South countries, particularly in Africa (see Cobbinah et al., 2015) has resulted in severe adverse effects on the environment, including pollution, water resource depletion, UHIs, sea-level rise, urban flooding, low agricultural activity, urban sprawl, slum development, deforestation, and wetland destruction, as well as myriad threats to public health (Abubakar & Aina, 2018). These adverse effects have had devastating impacts not only on humans, but also on other forms of life.

In exchange for paving an ideal path toward sustainable environmental management, the process of urbanization must be managed and appropriately controlled without surpassing the sustaining capacity of the environment. Creating solutions for rural development, which have the potential to lessen rural–urban migration, is one way to accomplish this goal (Baffoe et al., 2021). Moreover, other human activities that have a detrimental impact on the environment need to have their management improved, thus making it more effective and efficient. In addition, there is an urgent need to identify possible solutions by employing the organizational, legal, economic, administrative, and pedagogical tools that have been provided in this research. Developing and implementing strategies to mitigate the effects of urbanization requires cooperation between municipal agencies, the Ministry of Municipal and Rural Affairs, civic society, citizens, and international organizations.

Future research on this topic should provide frameworks and action plans for a sustainable urbanization strategy in Saudi Arabia. Given the potential for increasing urbanization to have a detrimental impact on human settlements, what steps can be taken to transform this trend into a constructive factor in sustainable development? How might the current UN framework for improving urban governance and expanding socio-economic transformation be improved? To lessen the likelihood of violent conflict occurring in an urban setting, how can social and cultural obstacles be resolved? Ultimately, considering everything that has hitherto been accomplished by the NUA, as well as the widespread recognition and promotion of Vision 2030, which aspects still require discussion to ensure sustainable growth through urbanization? These questions can be answered by future research.

Acknowledgements This chapter is based on the following previously published article and is reproduced with permission from Taylor & Francis: Almulhim, A. I., & Cobbinah, P. B. (2022). Urbanization-environment conundrum: An invitation to sustainable development in Saudi Arabian cities. *International Journal of Sustainable Development & World Ecology, 30*(4), 359–373.

References

Abubakar, I. R., & Aina, Y. A. (2018). Achieving sustainable cities in Saudi Arabia: Juggling the competing urbanization challenges. In N. Kock (Ed.), *E-planning and collaboration: Concepts, methodologies, tools, and applications* (pp. 234–255). IGI Global. https://doi.org/10.4018/978-1-5225-0187-9.ch003

Abubakar, I. R., & Bununu, Y. A. (2020). Low carbon city: Strategies and case studies. In *Sustainable cities and communities. Encyclopedia of sustainability in higher education* (pp. 366–376). Springer. https://doi.org/10.1007/978-3-319-95717-3_24

Abubakar, I. R., & Dano, U. L. (2020). Sustainable urban planning strategies for mitigating climate change in Saudi Arabia. *Environment, Development and Sustainability, 22*(6), 5129–5152. https://doi.org/10.1007/s10668-019-00417-1

ACAPS. (2014). *Secondary data review*. Retrieved on 20 May 2022 from https://www.acaps.org/sites/acaps/files/resources/files/secondary_data_review-sudden_onset_natural_disasters_may_2014.pdf. Accessed 13 June 2022.

Aina, Y. A. (2014). Urban sustainability perspectives on changes in adaptation to Saudi Arabian desert environment: The role of geospatial technologies. In G. M. Kumar & P. Moharana (Eds.), *Environment, people and development: Experiences from desert ecosystems* (pp. 331–348). NIPA.

Aina, Y. A. (2017). Achieving smart sustainable cities with GeoICT support: The Saudi evolving smart cities. *Cities, 71*, 49–58. https://doi.org/10.1016/j.cities.2017.07.007

Aina, Y. A., Merwe, J. H. V., & Alshuwaikhat, H. M. (2008). *Urban spatial growth and land use change in Riyadh: Comparing spectral angle mapping and band ratioing techniques*. In Proceedings of the academic track of the 2008 Free and Open Source Software for Geospatial (FOSS4G) Conference, incorporating the GISSA Conference, Cape Town, South Africa, pp. 51–57.

Al Shammary, S. (2020). *Environmental challenges in Saudi Arabia's oil and gas industry*. EcoMENA: Echoing Sustainability in MENA. https://www.ecomena.org/environmental-challenges-saudi-arabia-oil-and-gas-industry/. Accessed 13 June 2022.

Alatawi, A. S. (2022). Conservation action in Saudi Arabia: Challenges and opportunities. *Saudi Journal of Biological Sciences, 29*(5), 3466–3472. https://doi.org/10.1016/j.sjbs.2022.02.031

Aliu, I. R. (2022). Sustainable housing development dynamics in the global South: Reflections on theories, strategies and constraints. *Bulletin of Geography, Socio-Economic Series, 56*, 83–100. https://doi.org/10.12775/bgss-2022-0014

Al-Maamary, H. M. S., Kazem, H. A., & Chaichan, M. T. (2017). Climate change: The game changer in the Gulf Cooperation Council Region. *Renewable and Sustainable Energy Reviews, 76*, 555–576. https://doi.org/10.1016/j.rser.2017.03.048.

Almulhim, A. I. (2021). Public knowledge and perception of climate change and global warming in the context of environmental challenges and policies in Saudi Arabia. *WIT Transactions on Ecology and the Environment, 253*, 577–589.

Almulhim, A. I. (2022a). Understanding public awareness and attitudes toward renewable energy resources in Saudi Arabia. *Renewable Energy, 192*, 572–582. https://doi.org/10.1016/j.renene.2022.04.122

Almulhim, A. I. (2022b). Household's awareness and participation in sustainable electronic waste management practices in Saudi Arabia. *Ain Shams Engineering Journal, 13*(4), 101729. https://doi.org/10.1016/j.asej.2022.101729

Almulhim, A. I., & Abubakar, I. R. (2021). Understanding public environmental awareness and attitudes toward circular economy transition in Saudi Arabia. *Sustainability, 13*(18), 10157. https://doi.org/10.3390/su131810157

Almulhim, A. I., & Aina, Y. A. (2022). Understanding household water-use behavior and consumption patterns during COVID-19 lockdown in Saudi Arabia. *Water, 14*(3), 314. https://doi.org/10.3390/w14030314

Almulhim, A. I., Bibri, S. E., Sharifi, A., Ahmad, S., & Almatar, K. M. (2022). Emerging trends and knowledge structures of urbanization and environmental sustainability: A regional perspective. *Sustainability, 14*(20), 13195. https://doi.org/10.3390/su142013195

Al-Obaid, S., Samraoui, B., Thomas, J., El-Serehy, H. A., Alfarhan, A. H., Schneider, W., & O'Connell, M. (2017). An overview of wetlands of Saudi Arabia: Values, threats, and perspectives. *Ambio, 46*(1), 98–108. https://doi.org/10.1007/s13280-016-0807-4

Alqahtany, A., Rezgui, Y., & Li, H. (2014). A consensus-based framework for the sustainable urban planning development: As an approach for Saudi Arabian cities. *International Journal of Environmental Science and Development, 5*(2), 124–131. https://doi.org/10.7763/IJESD.2014.V5.463

Al-Subaiee, F. S. (2016). Socio-economic factors affecting the conservation of natural woodlands in Central Riyadh Area-Saudi Arabia. *Saudi Journal of Biological Sciences, 23*(3), 319–326. https://doi.org/10.1016/j.sjbs.2015.02.017

Alturki, F. K. (2015). *Promoting sustainable development through environmental law: Prospects for Saudi Arabia* [Master's thesis] [SJD Dissertation]. Pace University School of Law. https://digitalcommons.pace.edu/lawdissertations/17/. Accessed 13 June 2022.

Baffoe, G., Zhou, X., Moinuddin, M., Somanje, A. N., Kuriyama, A., Mohan, G., Saito, O., & Takeuchi, K. (2021). Urban–rural linkages: Effective solutions for achieving sustainable development in Ghana from an SDG interlinkage perspective. *Sustainability Science, 16*(4), 1341–1362. https://doi.org/10.1007/s11625-021-00929-8

Barbora. (2019). *How does poor waste management affect the climate change crisis.* GreenSail. https://www.green-sail.com/blog/how-does-poor-waste-management-affect-the-climate-change-crisis. posting by Barbora to Greensail Blog. Accessed 13 June 2022.

Baus, D. (2017). *Overpopulation and the impact on the environment* [Master's Thesis]. City University of New York. CUNY Academic Works. https://academicworks.cuny.edu/gc_etds/1906/. Accessed 13 June 2022.

Beckers, V., Poelmans, L., Van Rompaey, A., & Dendoncker, N. (2020). The impact of urbanization on agricultural dynamics: A case study in Belgium. *Journal of Land Use Science, 15*(5), 626–643. https://doi.org/10.1080/1747423X.2020.1769211

Belloumi, M., & Alshehry, A. S. (2016). The impact of urbanization on energy intensity in Saudi Arabia. *Sustainability, 8*(4), 375. https://doi.org/10.3390/su8040375

Bibri, S. E., Krogstie, J., & Kärrholm, M. (2020). Compact city planning and development: Emerging practices and strategies for achieving the goals of sustainability. *Developments in the Built Environment, 4*, 100021. https://doi.org/10.1016/j.dibe.2020.100021

Brauer, M., Freedman, G., Frostad, J., van Donkelaar, A., Martin, R. V., Dentener, F., van Dingenen, R., Estep, K., Amini, H., Apte, J.S., Balakrishnan, K., Barregard, L., Broday, D., Feigin, V., Ghosh, S., Hopke, P. K., Knibbs, L. D., Kokubo, Y., Liu, Y., … Cohen, A. (2016). Ambient air pollution exposure estimation for the global burden of disease 2013. *Environmental Science & Technology, 50*(1), 79–88. https://doi.org/10.1021/acs.est.5b03709

Byjus. (2020). *Meaning and functions of environment.* https://byjus.com/commerce/meaning-and-functions-of-environment/. Accessed 13 June 2022.

Chen, M., Zhang, H., Liu, W., & Zhang, W. (2014). The global pattern of urbanization and economic growth: Evidence from the last three decades. *PLoS ONE, 9*(8), e103799. https://doi.org/10.1371/journal.pone.0103799

Chowdhury, S., & Al-Zahrani, M. (2013). Implications of climate change on water resources in Saudi Arabia. *Arabian Journal for Science and Engineering, 38*(8), 1959–1971. https://doi.org/10.1007/s13369-013-0565-6

Cobbinah, P. B., Erdiaw-Kwasie, M. O., & Amoateng, P. (2015). Africa's urbanization: Implications for sustainable development. *Cities, 47*, 62–72.

Colenbrander, S. (2016). *Cities as engines of economic growth: The case for providing basic infrastructure and services in urban areas* [Working Paper]. International Institute for Environment and Development. https://pubs.iied.org/10801iied. Accessed 13 June 2022.

Dano, U., & Alqahtany, A. (2019). Issues undermining public transport utilization in Dammam city, Saudi Arabia: An expert-based analysis. *Journal of Sustainable Science and Management, 14*(2), 155–169.

Darfaoui, E. M., & Al Assiri, A. (2010). *Response to climate change in the Kingdom of Saudi Arabia.* Report prepared for Food and Agriculture Organization, United Nations.

Davies, R. (2021, April 28). *Saudi Arabia—Flash floods sweep through Mecca.* FloodList. https://floodlist.com/asia/saudi-arabia-floods-mecca-april-2021. Accessed 13 June 2022.

Dawson, R., Hall, J., Barr, S., Batty, M., Bristow, A., Carney, S., Dagoumas, A., Evans, S., Ford, A., Harwatt, H., Köhler, J., Tight, M., Walsh, C., & Zanni, A. (2017). A blueprint for the integrated assessment of climate change in cities. In *Green CITYnomics* (1st ed., pp. 32–51). Routledge.

DeNicola, E., Aburizaiza, O. S., Siddique, A., Khwaja, H., & Carpenter, D. O. (2015). Climate change and water scarcity: The case of Saudi Arabia. *Annals of Global Health, 81*(3), 342–353. https://doi.org/10.1016/j.aogh.2015.08.005

EIA. (2021). *Saudi Arabia.* https://www.eia.gov/international/analysis/country/SAU. Accessed 13 June 2022.

Elimam, H. (2022). Environmental problems and development sustainability in light of the kingdom's 2030 vision: Opportunities and challenges. *International Journal of Education and Social Science, 9*(1), 2410–5171.

Evans, M. (2020, July 7). *What is environmental sustainability?* The Balance Small Business. https://www.thebalancesmb.com/what-is-sustainability-3157876. Accessed 13 June 2022.

Fang, C., & Yu, D. (2017). Urban agglomeration: An evolving concept of an emerging phenomenon. *Landscape and Urban Planning, 162*, 126–136. https://doi.org/10.1016/j.landurbplan.2017.02.014

Freudenrich, C. (2022). *How oil refining works.* howstuffworks. https://science.howstuffworks.com/environmental/energy/oil-refining1.htm. Accessed 13 June 2022.

Galli, A., Iha, K., Moreno Pires, S., Mancini, M. S., Alves, A., Zokai, G., Lin, D., Murthy, A., & Wackernagel, M. (2020). Assessing the ecological footprint and biocapacity of Portuguese cities: Critical results for environmental awareness and local management. *Cities, 96*, 102442. https://doi.org/10.1016/j.cities.2019.102442

Gazzeh, K., & Abubakar, I. R. (2018). Regional disparity in access to basic public services in Saudi Arabia: A sustainability challenge. *Utilities Policy, 52*, 70–80. https://doi.org/10.1016/j.jup.2018.04.008

General Authority for Statistics. (2022). *General Authority for Statistics—Saudi Arabia.* https://www.stats.gov.sa/en. Accessed 13 June 2022.

gov.sa. (2021). *Environmental protection in KSA.* https://www.my.gov.sa/wps/portal/snp/aboutksa/environmentalProtection/!ut/p/z0/04_Sj9CPykssy0xPLMnMz0vMAfIjo8zivQIsTAwdDQz9LSw8XQ0CnT0s3JxDfA0M_A30g1Pz9AuyHRUBPXGyKQ!!/. Accessed 13 June 2022.

Im, H. D. (2012). *Saudi Arabia. Essay of Saudi Arabia—Three environmental issues and IT technology.* https://www.ics.uci.edu/~wmt/courses/ICS5_W13/SaudiArabia.html. Accessed 13 June 2022.

International Trade Administration. (2022). *Saudi Arabia—Country commercial guide.* https://www.trade.gov/country-commercial-guides/saudi-arabia-waste-management. Accessed 13 June 2022.

IPCC. (2007). Impacts, adaptation, and vulnerability. In M. Parry, O. Canziani, J. Palutikof, P. van der Linden, & C. Hanson (Eds.), *Climate change contribution of Working Group II to the Fourth Assessment Report of the Intergovernmental Panel on Climate Change.*

Jianping, L., Minrong, L., Jinnan, W., Jianjian, L., Hongwen, S., & Maoxing, H. (2014). Global environmental issues and human wellbeing. In *Report on global environmental competitiveness* (pp. 3–21). Springer. https://doi.org/10.1007/978-3-642-54678-5_1

Kadi, M. W. (2009). Soil pollution hazardous to environment: A case study on the chemical composition and correlation to automobile traffic of the roadside soil of Jeddah city, Saudi Arabia. *Journal of Hazardous Materials, 168*(2–3), 1280–1283. https://doi.org/10.1016/j.jhazmat.2009.03.015

Kennedy, G. (2009). *Food security in the context of urban sub-Saharan Africa.* Internet paper for food security, food Africa. Internet Forum, 31 March 2011, April 2003.

Kuddus, M. A., Tynan, E., & McBryde, E. (2020). Urbanization: A problem for the rich and the poor? *Public Health Reviews, 41*(1), 1. https://doi.org/10.1186/s40985-019-0116-0

Koroso, N., Lengoiboni, M., & Zevenbergen, J. (2021). Urbanization and urban land use efficiency: Evidence from regional and Addis Ababa satellite cities, Ethiopia. *Habitat International, 117*, 102437. https://doi.org/10.1016/j.habitatint.2021.102437

Li, X., & Zhou, Y. M. (2017). Offshoring pollution while offshoring production? *Strategic Management Journal, 38*(11), 2310–2329. https://doi.org/10.1002/smj.2656

Liu, Y., Huang, X., Yang, H., & Zhong, T. (2014). Environmental effects of land-use/cover change caused by urbanization and policies in Southwest China Karst area—A case study of Guiyang. *Habitat International, 44*, 339–348. https://doi.org/10.1016/j.habitatint.2014.07.009

Mahmood, H., Alkhateeb, T. T. Y., Al-Qahtani, M. M. Z., Allam, Z., Ahmad, N., & Furqan, M. (2019). Agriculture development and CO_2 emissions nexus in Saudi Arabia. *PLoS ONE, 14*(12), e0225865. https://doi.org/10.1371/journal.pone.0225865

Mahmood, H., Alkhateeb, T. T. Y., & Furqan, M. (2020). Industrialization, urbanization and CO_2 emissions in Saudi Arabia: Asymmetry analysis. *Energy Reports, 6*, 1553–1560. https://doi.org/10.1016/j.egyr.2020.06.004

Mason, D., Iida, A., Watanabe, S., Jackson, L. P., & Yokohari, M. (2020). How urbanization enhanced exposure to climate risks in the Pacific: A case study in the Republic of Palau. *Environmental Research Letters, 15*(11), 114007. https://doi.org/10.1088/1748-9326/abb9dc

Migrant Refugees. (2020, December). *Migration profile: Saudi Arabia.* https://migrants-refugees.va/it/wp-content/uploads/sites/3/2021/07/2020-CP-Saudi-Arabia.pdf. Accessed 13 June 2022.

Mohajerani, A., Bakaric, J., & Jeffrey-Bailey, T. (2017). The urban heat island effect, its causes, and mitigation, with reference to the thermal properties of asphalt concrete. *Journal of Environmental Management, 197*, 522–538. https://doi.org/10.1016/j.jenvman.2017.03.095

Mutuku, C. (2018). *Urbanization in developing countries.* GRIN Verlag. https://www.grin.com/document/432475. Accessed 13 June 2022.

Nasrudin, A. (2022). *Sociocultural environment: Meaning, variables, impact on the business.* Penpoin. https://penpoin.com/sociocultural-environment/. Accessed 13 June 2022.

Nassar, D. M., & Elsayed, H. G. (2018). From informal settlements to sustainable communities. *Alexandria Engineering Journal, 57*(4), 2367–2376. https://doi.org/10.1016/j.aej.2017.09.004

National Astronautics and Space Administration [NASA]. (2022.) *Sea level.* https://climate.nasa.gov/vital-signs/sea-level/. Accessed 13 June 2022.

National Audit Office. (2015). *A short guide to environmental protection and sustainable development.* https://www.nao.org.uk/wp-content/uploads/2015/09/A-Short-Guide-to-Environmental-protection-and-sustainable-development.pdf. Accessed 13 June 2022.

Odhiambo, G. O. (2017). Water scarcity in the Arabian Peninsula and socio-economic implications. *Applied Water Science, 7*(5), 2479–2492. https://doi.org/10.1007/s13201-016-0440-1

Revi, A., Satterthwaite, D., Aragón-Durand, F., Corfee-Morlot, J., Kiunsi, R. B., Pelling, M., & Sverdlik, A. (2014). Towards transformative adaptation in cities: The IPCC's fifth assessment. *Environment and Urbanization, 26*(1), 11–28. https://doi.org/10.1177/0956247814523539

Salama, A. M., & Alshuwaikhat, H. (2006). A transdisciplinary approach for a comprehensive understanding of sustainable affordable housing. *Global Built Environment Review, 5*(3), 35–50.

Santamouris, M. (2013). Using cool pavements as a mitigation strategy to fight urban heat island—A review of the actual developments. *Renewable and Sustainable Energy Reviews, 26*, 224–240. https://doi.org/10.1016/j.rser.2013.05.047

Saudi Green Initiative. (2022). *Saudi Green Initiative aims to improve quality of life.* https://www.saudigreeninitiative.org/about-saudi-green-initiative/. Accessed 13 June 2022.

Shaban, A., Kourtit, K., & Nijkamp, P. (2022). Causality between urbanization and economic growth: Evidence from the Indian states. *Frontiers in Sustainable Cities, 4*, 901346. https://doi.org/10.3389/frsc.2022.901346

Sharkh, B. A., Al-Amoudi, A. A., Farooque, M., Fellows, C. M., Ihm, S., Lee, S., & Voutchkov, N. (2022). Seawater desalination concentrate—A new frontier for sustainable mining of valuable minerals. *npj Clean Water, 5*(1), 1–16.

Shava, G. N., Hleza, S., Tlou, F., Shonhiwa, S., & Mathonsi, E. (2021). Qualitative content analysis utility, usability and processes in educational research. *International Journal of Research and Innovation in Social Science (IJRISS), 5*(7), 553–558.

The World Bank. (2018). *3 big ideas to achieve sustainable cities and communities*. https://www.worldbank.org/en/news/immersive-story/2018/01/31/3-big-ideas-to-achieve-sustainable-cities-and-communities. Accessed 13 June 2022.

The World Bank. (2020). *Urban development overview*. https://www.worldbank.org/en/topic/urbandevelopment/overview. Accessed 13 June 2022.

The World Bank. (2022a). *Saudi Arabia GDP*. https://tradingeconomics.com/saudi-arabia/gdp. Accessed 13 June 2022.

The World Bank. (2022b). *World Development Indicators*. https://datacatalog.worldbank.org/search/dataset/0037712. Accessed 13 June 2022.

Tolba, M. K., & Saab, N. W. (2009). *Arab environment climate change* (pp. 7–8). Arab Forum for Environment and Development (AFED) with Technical Publications and Environment and Development Magazine.

UN-Habitat. (2011). *Global report on human settlements 2011 cities and climate change*. Earthscan Publications. https://unhabitat.org/global-report-on-human-settlements-2011-cities-and-climate-change. Accessed 13 June 2022.

UN-Habitat. (2019). *Saudi Cities Report 2019. Future Saudi cities program*. https://unhabitat.org/sites/default/files/2020/05/saudi_city_report.english.pdf. Accessed 13 June 2022.

UN-Habitat. (2020). *World Cities Report 2020. The value of sustainable urbanization*. https://unhabitat.org/sites/default/files/2020/10/wcr_2020_report.pdf

UN-Habitat. (2022). *World Cities Report 2022. Envisaging the future of cities*. https://unhabitat.org/sites/default/files/2022/06/wcr_2022.pdf. Accessed 13 June 2022.

United Nations [UN]. (2021). *Make cities and human settlements inclusive, safe, resilient and sustainable*. https://sdgs.un.org/goals/goal11

United States Energy Information Administration. (2000). *Saudi Arabia: Environmental issues*. Albawaba. https://www.albawaba.com/business/saudi-arabia-environmental-issues-part-one. Accessed 13 June 2022.

US EPA. (2021). *Learn about heat islands*. https://www.epa.gov/heatislands/learn-about-heat-islands. Accessed 13 June 2022.

U.S Geological Survey. (2003). *Groundwater depletion across the nation*. Retrieved from https://pubs.usgs.gov/fs/fs-103-03/#:~:text=Some%20of%20the%20negative%20effects,with%20any%20ground%2Dwater%20use. Accessed 13 June 2022.

Wan, G. (2013). Why urbanization may benefit the environment. *Business Times Singapore*. https://www.adb.org/news/op-ed/why-urbanization-may-benefit-environment-guanghua-wan#:~:text=Urban%20living%20encourages%20walking%20and,operate%20in%20an%20urban%20setting. Accessed 13 June 2022.

Weiskopf, S. R., Rubenstein, M. A., Crozier, L. G., Gaichas, S., Griffis, R., Halofsky, J. E., Hyde, K. J. W., Morelli, T. L., Morisette, J. T., Muñoz, R. C., Pershing, A. J., Peterson, D. L., Poudel, R., Staudinger, M. D., Sutton-Grier, A. E., Thompson, L., Vose, J., Weltzin, J., & Whyte, K. P. (2020). Climate change effects on biodiversity, ecosystems, ecosystem services, and natural resource management in the United States. *Science of the Total Environment, 733*, 137782–137782. https://doi.org/10.1016/j.scitotenv.2020.137782

Wilson, B., & Chakraborty, A. (2013). The environmental impacts of sprawl: Emergent themes from the past decade of planning research. *Sustainability, 5*(8), 3302–3327. https://doi.org/10.3390/su5083302

World Economic Forum. (2022). *This chart shows the impact rising urbanization will have on the world*. https://www.weforum.org/agenda/2022/04/global-urbanization-material-consumption/. Accessed 13 June 2022.

World Health Organization [WHO]. (2021, October 29). *Urban health*. https://www.who.int/news-room/fact-sheets/detail/urban-health. Accessed 13 June 2022.

World Population Review. (2022). *Saudi Arabia population 2022*. https://worldpopulationreview.com/countries/saudi-arabia-population. Accessed 13 June 2022.

Worldometer. (2022). *Saudi Arabia population*. https://www.worldometers.info/world-population/saudi-arabia-population/. Accessed 13 June 2022.

Zafar, S. (2021). *Municipal waste management in Saudi Arabia*. Bioenergy Consult. https://www.bioenergyconsult.com/municipal-wastes-in-saudi-arabia

Zhang, X. (2016). The trends, promises and challenges of urbanisation in the world. *Habitat International, 54*, 241–252. https://doi.org/10.1016/j.habitatint.2015.11.018

Part II
Planning and Resilience in Practice

Chapter 4
Mediating Urbanization and Sustainable Development in Saudi Arabia: Urban Planning Pivotality and Stewardship

4.1 Introduction

Historical and contemporary accounts of urbanization present both opportunities (e.g., infrastructure provision) and challenges (e.g., urban sprawl) for cities globally (Almulhim & Cobbinah, 2024a; asghar Pilehvar, 2021; Bai et al., 2017; Chen et al., 2014; The World Bank, 2020). Managing these complications is important if we are to transition toward urban sustainability.[1] What these complications are, and their entanglements with urban sustainability is deeply foregrounded when emphasis is placed on global south cities commonly faced with urban planning inadequacies (Cobbinah et al., 2015; Mutuku, 2018). Traditional arguments indicate that rapid urbanization is routinely attributed to the common belief that urban population expansion via rural–urban transfer and natural increase intrinsically provides opportunities for socio-economic growth and production efficiency, thus contributing to urban sustainability (see Shaban et al., 2022; The World Bank, 2020; UN-Habitat, 2019a). However, mounting evidence has shown that this is not always the case as urbanization processes across global south cities is threatening urban sustainability (Almulhim et al., 2024; Cobbinah et al., 2015; Elmqvist et al., 2021; Kahouli et al., 2022; World Health Organization, 2021) and producing socio-economic deterioration and natural resource degradation (Geneva Environment Network, 2022; Muhammad, 2021).

Urban planning, as a distinguished mechanism for urban sustainability, seeks improvements in overall urban living environment, ranging from spatial, economic, environmental, and social conditions (Abubakar & Dano, 2020). As Archistar (2024, p. 1) explained, "with a thoroughly planned layout, a city can provide its residents with access to all essential services, points of interest and amenities. At the same time, the

[1] Urban sustainability is explained as "an adaptive process of facilitating and maintaining a virtual cycle between ecosystem services and human wellbeing through concerted ecological, economic, and social actions in response to changes within and beyond the urban landscape" (Wu, 2014, p. 213).

A. I. Almulhim and P. B. Cobbinah, *Urbanization Rethought*, Urban Sustainability,
https://doi.org/10.1007/978-981-92-0118-1_4

unfavourable aspects of urban life are reduced, leading to an overall healthier lifestyle and improved quality of life." Within this context, urban planning interventions often focus on reducing inequality, addressing infrastructural deficits, poverty reduction, and enhancing urban livability and well-being (Abubakar & Aina, 2018; Almulhim & Cobbinah, 2023a), in addition to urban sustainability (Cobbinah & Darkwah, 2017; Wu, 2014), and climate change management (Cobbinah et al., 2019). Since the 1980s, such urban planning functions have been strongly associated with the idea of urban sustainability, centered on actions and initiatives that produce socio-economic and environmental progress for addressing intragenerational and intergenerational inequity, social and spatial injustice (see Mersal, 2016; Naess, 2001). Fundamentally, these outcomes of effective urban planning reflect a continuous and positive change in the urban space where the negative externalities of urbanization are controlled, leading to sustainable urbanization (Abubakar & Aina, 2018; Almulhim & Cobbinah, 2023a).

However, such an approach to urban planning is often obscured in global south cities where efforts that promote equity, control sprawl, reduce poverty, and spur ethical environmental practices are inadequate and impotent amidst rapid urbanization (Cobbinah, 2017; Fuseini & Kemp, 2015). Considering the challenges associated with rapid urbanization in global south cities (see for example, UN-Habitat, 2020), ways in which urban planning-led development can promote sustainable urbanization outcomes must be explored and understood. Yet, while research on urbanization and urban planning is mushrooming in major geopolitical regions of the global south including Asia and Africa, an understanding of urbanization processes and their association with urban planning is now emerging in the Middle East. This chapter seeks to better understand how urban planning-led development can contribute to sustainable urbanization outcomes in Saudi Arabian cities by addressing two important research questions: (i) In what ways do urbanization and urban planning intersect in urban development process? and (ii) What are the barriers, drivers, processes, and outcomes associated with urban planning-led development?

Framed around sustainable development thinking, a case study research approach (Yin, 2015) involving three major cities in Saudi Arabia was used to offer two important contributions to urban studies and planning literature. First, the chapter addresses a current gap in the literature by providing an understanding of the necessary conditions and subsequent processes of urbanization, and the criticality of urban planning in shaping sustainable urbanization. And second, the chapter examines urban planning-urbanization intersections and subsequent processes leading to sustainable urbanization outcomes. Saudi Arabian cities were analyzed to realize the research purpose because they are experiencing rapid urbanization and contending urban planning inadequacies. This chapter is organized into five parts. Following the introduction section is the literature review section analyzing urbanization within the context of sustainable development with a focus on Saudi Arabia. Section 4.3 discusses the methodology for this research while Sect. 4.5 presents the results and discussion. Finally, the conclusion is presented in Sect. 4.6.

4.2 Urbanization-Sustainable Development Linkages: A Focus on Saudi Arabia

The purpose of this section is, first, to establish the relations between urbanization and sustainable development, and second to demonstrate the positionality of Saudi Arabia in the urbanization discourse. The background to and need for sustainable urban development is well cataloged (see for example Rasoolimanesh et al., 2012; Roy, 2009; Shen et al., 2011; UN-Habitat/DFID, 2002; Whitehead, 2003). However, a relatively recent discourse connecting urbanization processes with sustainable development agendas has emerged (Mtapuri & Myeni, 2019; Solly et al., 2021; Tan et al., 2016). For example, the sustainable development goal (SDG) 11 focuses on sustainable urbanization emphasizing creation of sustainable, resilient, and inclusive cities, while the New Urban Agenda provides an incisive, action-focused, future-oriented, and comprehensive framework of initiatives for sustainable urban development. Accordingly, sustainable urbanization has gained the attention of urban studies and planning scholars due to its potential to address the negative complications of rapid urban growth (cf. Almulhim et al., 2024; Rasoolimanesh et al., 2012; Whitehead, 2003). Nonetheless, several researchers have warned that urbanization management per se does not inherently lead to sustainable development (Malkina-Pykh, 2002; Roy, 2009). This is the case, for instance, when existing and ongoing urbanization management agendas are uncoordinated (e.g., between formal and informal communities) and are not urban planning guided (Rasoolimanesh et al., 2012). Thus, a need arises to move forward the discussion of urbanization management to sustainable urbanization founded on urban planning-led agenda. In here, it is argued that such a purpose can be realized by developing a deeper insight of urbanization dynamics and processes, especially in global south countries.

Fundamental to such insight is the sustainable development theory (Mensah, 2019; Wu, 2010). Unlike other theories of urban development (see Roser (2023), on economic growth and Black and Henderson (1999), on urban growth), sustainable development theory, according to the Brundtland Commission Report (Schaefer & Crane, 2005), views urban development as a process that addresses the needs and problems of the present urban inhabitants without compromising the ability of the next generations to meet their own needs. In this process, urban development can occur and interact with the environment without damaging or risking the resource for the future (Mensah, 2019). From a sustainable development perspective, urbanization is a pervasive feature of urban development through which socio-economic and environmental needs of society can be fulfilled.

As presented in Table 4.1, rapid urbanization is a defining feature of all major Saudi Arabian cities with the country recognized as one of the world's most urbanized as approximately 80% of residents live in metropolitan areas (UNDRR, 2018). The nation witnessed an annual population has increased of 2.5% since 2015, with the population surging from 16 million in 1990 to its current population of approximately 34 million, more than doubling in the last quarter-century. Two notable demographic shifts are the nation's aging population and a decline in the proportion of children and those of working age. The share of those over 65 years rose from 2.6% in 1985 to 3.1% in 2015, with forecasts indicating an increase to 6.6% by 2030 and 16.7% by 2050 (UN-Habitat, 2019a). This aging trend necessitates that Saudi cities adapt to meet the requirements of older adults in areas such as transport, recreation, and healthcare. Sociocultural changes, including the evolving role of women due to modernization and education, are leading to a decrease in the numbers of children, adolescents, and working-age populations. The number of children has dwindled with the shift from a rural to an urban lifestyle (Kuncic & Andrés, 2021).

The five largest urban agglomerations in the country, Riyadh, Jeddah, Makkah, Al-Madinah, and Dammam (see Fig. 4.1), are major contributors to this growth. Riyadh, the capital city, is expected to grow from 6,218,000 residents in 2015 to 9,058,000 by 2035. Jeddah, the second-largest city, will also see substantial growth, from 4,035,000 in 2015 to an anticipated 5,710,000 by 2035 (UN-Habitat, 2019a). Saudi Arabia's urbanization has produced major territorial and local disparities, with the larger cities relatively well positioned to benefit from urbanization in terms of infrastructure provision and socio-economic opportunities (UN-Habitat, 2019a). With approximately 83% of Saudi Arabia's populace residing in urban areas (UN-Habitat, 2019a), urbanization has contributed to improved healthcare access, economic prospects, and overall quality of life. In this case, urbanization can be a revolutionary driver for sustainable development when planned and controlled.

Despite Saudi Arabia's growth being guided by a spatial strategy aimed at achieving a comprehensive approach across all areas, as well as a fairer population dispersion, the dynamics of urban development have distinctly favored specific major urban centers. These centers have emerged as the primary beneficiaries of

Table 4.1 Patterns of urbanization in major Saudi Arabian cities (2015–2035 projections)

Year (in thousands)	2015 ('000)	2020 ('000)	2025 ('000)	2030 ('000)	2035 ('000)
Total population	31,557	34,710	37,290	39,480	41,317
Urban population	26,249	29,256	31,843	34,143	36,170
Level of urbanization (%)	83.2	84.3	85.4	86.5	87.5
Five largest urban agglomerations					
Riyadh	6218	7231	7953	8547	9058
Jeddah	4035	4610	5022	5388	5710
Makkah	1796	2042	2219	2379	2521
Al-Madinah	1299	1489	1625	1744	1848
Dammam	1080	1253	1376	1478	1566
Total population of the five largest agglomerations	14,428	16,625	18,195	19,536	20,703
Five largest agglomerations as % of urban population	54.97	56.82	57.14	57.22	57.24
Five largest agglomerations as % of total population	45.72	47.90	48.79	49.48	50.12

Source Authors' compilation from UN-Habitat (2019a)

Fig. 4.1 Major urban agglomerations in Saudi Arabia and transportation network adapted and modified from UN-Habitat (2019a)

the economic prosperity driven by oil revenues and increased productivity (UN-Habitat, 2019a), highlighting the challenges and opportunities for urban planning, and sustainable urbanization (UN-Habitat, 2019a). The increasing concentration of population in cities coupled with the many advantages, such as better facilities and infrastructure that boost output, increase job prospects, and raise standard of living, has led to the surrounding territories' abandonment and an inability to successfully harness and use their natural riches. For example, Riyadh—the national capital—has benefited significantly from initiatives pertaining to urban planning and growth, large-scale foreign direct investments, first-rate amenities, and an influx of foreign workers (Al Bassam, 2012; UNDP, 2022). These factors have strengthened Riyadh's standing in Saudi Arabia's settlement hierarchy and elevated it to the fifth-largest land-area city in the Middle East (Abubakar & Aina, 2018). Heightened urban dominance not only sidelines smaller and medium-sized cities in the national development agenda but also perpetuates inequalities and disrupts the country's urban framework (Almulhim & Cobbinah, 2023a).

In Saudi Arabia, urban primacy has arisen due to a combination of factors: the increasing size and spatial concentration of urban populations, economic progress, an unbalanced investment approach favoring specific cities, the shift of rural zones into urbanized areas, global migration, and population expansion attributed to the movement of people from rural-to-urban localities (UN-Habitat, 2019a). In this case, urbanization can be a catalyst for economic growth but requires effective urban planning to deliver sustainable outcomes. The World Bank researchers have highlighted essential factors which border on urban planning and management, including land use, a viable land market providing enough land for housing, strong leadership across different levels of governance, the enforcement of zoning laws, efficient tax collection, investment in public infrastructure, private capital investment, and the formation of housing capital both locally and nationally (Almulhim & Cobbinah, 2023a; World Bank, 2020). A successful use of urbanization in Saudi Arabia thus primarily depends on strong urban planning across different levels of administration (Almulhim & Cobbinah, 2023a). It is within the foregoing context that this study is framed to provide an understanding of the intersection of urban planning and urbanization in shaping the future of Saudi Arabian cities.

4.3 Research Approach

4.3.1 Study Area Context

This study focuses on Saudi Arabian cities. Saudi Arabia, spanning the greater part of the Arabian Peninsula and situated at the meeting point of three continents, Africa, Europe, and Asia, ranks as the second-largest Arab nation. This vast country covers approximately 2.15 million square kilometers, making it the largest in the Middle East and the 13th largest globally (Almulhim & Cobbinah, 2024b). Saudi Arabia's population has increased by 2.5% annually since 2015. As previously mentioned in Sect. 2, over the past 25 years, the country's population has more than doubled, increasing dramatically from 16 million in 1990 to over 32 million in 2023. By 2050, the population is expected to increase by 13.5 million from 2015 levels or, if the pace of growth continues, 45 million people (UN-Habitat, 2019a). Similarly, urban sprawl (Alharthi & El-Damaty, 2022), excessive energy usage (Kahouli et al., 2022), inadequate infrastructure and essential services, and environmental degradation and traffic congestion are among the major urbanization-induced challenges confronting the country (Alhowaish, 2015; Kahouli et al., 2022).

Three case study cities including Riyadh, Jeddah, and Dammam were selected to further highlight the complexity and diversity of the phenomenon of sustainable urbanization. The selection was based on their geographical diversity covering central (Riyadh), western (Jeddah), and eastern (Dammam) parts of the country, as well as their urbanization dynamics. As presented in Table 4.1, Riyadh has a population of over 7 million (most populated urban agglomeration), while Jeddah has over 4 million (medium-sized urban agglomeration) and Dammam has over a million residents (small-sized urban agglomeration). These cases provide an understanding of the complexity of urbanization and represent the urban development narrative in Saudi Arabia.

The emergence and evolution of urban planning in Saudi Arabia is classified into two major periods (Al-Hathloul & Anis-ur-Rahmaan, 1985; Mubarak, 2004). First, urban planning during the period between the late 1930s and early 1970s focused on everyday needs of urbanites emphasizing on physical planning mostly master plan. The second period starting the 1970s shifted focus toward comprehensive urban planning with interdisciplinary emphasis starting from Five Year National Plans to Regional and Master Development Plans (Al-Hathloul & Anis-ur-Rahmaan, 1985; Garba, 2004; Mandeli, 2008). These periods were further characterized by the introduction of legislation and capacity training for urban planning. Although the comprehensive urban planning process in Saudi Arabia is described as ambitious and dynamic compared to other regions of the world, the ongoing urbanization challenges have limited urban planning's influence on achieving sustainable urbanization (Al-Hathloul & Mughal, 2004). Notwithstanding these developments, urban planning has not been inadequate in promoting sustainable urbanization. Although progressive in its objectives, the ambitious planning process in Saudi Arabia has frequently been impeded by fast population increase, environmental deterioration, and insufficient local policy implementation (Al-Hathloul & Mughal, 2004; Alotaibi & Potoglou, 2018). Moreover, the centralized government structure in Saudi Arabia has occasionally resulted in a mismatch between planning attempts at the national level and the actual conditions in local cities (Aina et al., 2019). During this time, oil income has significantly influenced the determination of urban development, which frequently prioritized economic expansion over environmentally sustainable urban practices (Garba, 2004; Mubarak, 2004).

The modern metropolitan environment in Saudi Arabia continues to be shaped by the repercussions of urban planning inadequacies. For example, problems like urban sprawl and environmental degradation are a result of the early emphasis on physical planning without enough integration of social and environmental considerations (Mandeli, 2008). Meanwhile, the constraints of centralized governance institutions and the difficulties of rapidly urbanizing populations have limited the comprehensive planning initiatives of the 1970s and beyond, despite their greater holistic breadth (Aina et al., 2019).

To have a comprehensive understanding of Saudi Arabia's urban growth phenomena, a deeper investigation into these historical phases is essential. The persistent obstacles associated with urbanization, such as urban sprawl, environmental sustainability, and social equality, underscore the constraints of these antiquated methodologies and the imperative for perpetual advancement in urban planning techniques (Mandeli, 2008). This background lays the groundwork for a thorough examination of Saudi Arabia's sustainable urbanization, especially when viewed through the prism of the three case study cities, where the interaction between historical planning initiatives and contemporary urban problems is increasingly evident (Garba, 2004). This provides a context to explore sustainable urbanization and the influence of urban planning in Saudi Arabia, using the three case study cities.

4.3.2 Data Collection and Analysis

The study was conducted using Preferred Reporting Items for Systematic Reviews and Meta-Analyses (PRISMA) framework. We reviewed both conceptual and empirical studies on urban planning in the global South, focusing on Saudi Arabia. PRISMA framework is a useful for methodically conducting transparent and replicable procedures for consolidating existing research findings on a particular topic (Page et al., 2021). In this chapter, we used secondary data to explore the framings of urban planning and provide critical analysis of sustainable urbanization by examining the barriers, drivers, processes, and outcomes associated with urban planning-led development in Saudi Arabian cities. Using secondary sources, data were sourced from major databases such as Scopus and Web of Science covering grey literature, including statistical data and technical reports from institutional archives, websites of international developmental organizations (e.g., UNDRR, UNDP, UN-Habitat), and governmental bodies (e.g., Ministry of Municipality and Rural Affairs).

From August to October 2023, a search query (Table 4.2) was conducted using the following keywords: "urban sprawl," "sustainable development," "city management," "infrastructural development," "urban governance," "urbanization," "environmental pollution," "climate change adaptation," "urban resilience," "urban planning," and "population growth." These keywords served to refine the search parameters and establish the boundaries of the study. Selection criteria were applied to ensure the relevance and quality of the literature, which included full-text journal books, book chapters, articles, and policy documents—published in English with a contextual focus on Saudi Arabia, and the start and end year of 1985 and 2023 respectively, and were relevant to the study objectives.

A total of 9489 records were generated from the literature search across the two databases (Scopus = 5687 and WoS = 3802). After duplication checks, 7325 records were removed. Additional checks were carried out with their titles and keywords to establish their relevance to the chapter, resulting in the elimination of 8328 records. The remaining list of records ($n = 351$) was then assessed against the inclusion and exclusion criteria by focusing individually on the title and abstracts of the records (Table 4.3). This led to the removal of 268 records, remaining 83 records for the final selection and inclusion (Fig. 4.2). The final stage entailed structuring and examining the amassed data through content analysis. Data sets and scholarly works were sorted

Table 4.2 Search strategy

S. No	Databases	Number of records	Search query
1	Scopus	5687	TITLE-ABS-KEY ("urban sprawl*" OR "sustainable development" OR "city management*" OR "infrastructural development*" OR "urban governance*" OR "urbanization*" OR "environmental pollution*" OR "climate change adaptation*" OR "urban resilience*" OR "urban planning*" OR"population growth") AND AFFILCOUNTRY ("Saudi Arabia") AND NOT (2024) AND (EXCLUDE (PUBYEAR, 1979) OR EXCLUDE (PUBYEAR, 1982) OR EXCLUDE (PUBYEAR, 1984)) AND (LIMIT-TO (DOCTYPE, "ar") OR LIMIT-TO (DOCTYPE, "cp") OR LIMIT-TO (DOCTYPE, "re") OR LIMIT-TO (DOCTYPE, "ch") OR LIMIT-TO (DOCTYPE, "bk")) AND (LIMIT-TO (LANGUAGE, "English"))
2	Web of Science 2000–2023	3802	TS = (("urban sprawl*" OR "sustainable development" OR "city management*" OR "infrastructural development*" OR "urban governance*" OR "urbanization*" OR "environmental pollution*" OR "climate change adaptation*" OR "urban resilience*" OR "urban planning*" OR "population growth")) AND CU = ("Saudi Arabia")
	Total	**9489**	

by related topics, with each document meticulously assessed to distill themes pertinent to the research goals. Subsequently, these themes were compiled, integrated, and brought into alignment. This research method offered an efficient strategy for collecting and evaluating relevant data for this analysis.

Table 4.3 Inclusion and exclusion publication criteria

Inclusion criteria	Exclusion criteria
Climate change	Irrelevant to research aim and objectives
Urban population growth	Rural planning and development
Urban sprawl	Irrelevant to global South-focused urban planning
Urbanization	Not applicable to the Arabian context
Urban governance	
Sustainable development	
City management	
Urban planning in Saudi Arabia	
Urban land use	
Infrastructural development	
Relevant to the research purpose	

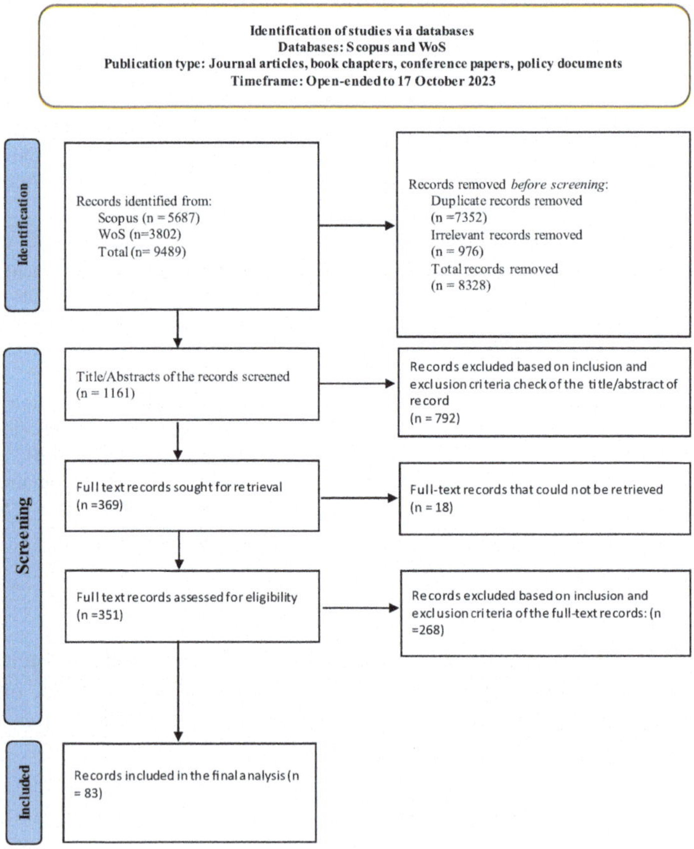

Fig. 4.2 PRISMA framework for analysis

4.4 Urban Planning and Sustainable Urbanization Narrative in Saudi Arabia

Our analysis of the three case study cities shows unprecedented levels of urbanization reflected in both population growth and spatial extent of cities. On population growth, Riyadh is the largest city with an annual growth rate of 4% which is higher than the national average of 2.11% (see Table 4.4). Similarly, Jeddah, serving as the economic hub and port of western part of Saudi Arabia, and Dammam, as an Arabian gulf city, have witnessed significant growth in their population (see Table 4.4), with an average annual growth rate of 3.2% and 3% respectively. Available spatial data indicate that this growth in population is also reflected in the spatial expansion of these cities. Our analysis indicates that urban expansion in Riyadh began in the 1960s in response to the oil boom, with significant expansion occurring in the 1980s. Since this period, Riyadh has nearly doubled in both population and urban expansion every two decades (UN-Habitat, 2019d). As a result, densities have decreased and land allocation per capita has increased, putting a pressure on the city's land resource and resulting in an unsustainable growth and development pattern.

Jeddah, with a population of 4,082,184, covers a land area of 84,675 hectares. While the density analysis indicates variations in density, over 610,000 residents representing 15% of Jeddah's population reside in areas with high density of more than 300 p/ha. Conversely, about 18,930 hectares were occupied by 1,681,000 residents, representing 42% of total population of Jeddah, demonstrating a medium to low density of 50–150 p/ha. Overall, our analysis shows that Jeddah's spatial extent and population have increased by over 400% and 1000% respectively between 1970 and 2017 (see Fig. 4.3). In Dammam, most spatial development occurred at the urban core with the highest densities from 80 to 120 p/ha with the average density of 20.13 p/ha (Ministry of Municipality & Rural Affairs, 2007; UN-Habitat, 2019b) (see Fig. 4.3). As a result, Dammam has a decreasing density due to increasing land speculation driving urban development with urban expansion from 2096 hectares in 1973 to 20,368 hectares in 2015, doubling 10 times in four decades.

Our analysis across the three case study cities shows that urbanization is rapid facilitated by industrialization and modernization and has led to a reduction in the effectiveness of various institutions in Saudi Arabian urban centers. Yet, urban planning has not demonstrated significant capacity in managing urbanization. The

Table 4.4 Urbanization information for the case studies

City	Population	Population density on built-up area	Age profile	Population growth rate (%)
Riyadh	6,506,700	71 p/ha	58% < 30	4
Jeddah	4,082,184	41.21 p/ha	41% < 24	3.2
Dammam	2,250,000	20.13 p/ha	50% < 30	3

Source Authors' compilation from UN-Habitat (2019b, 2019c, 2019d)

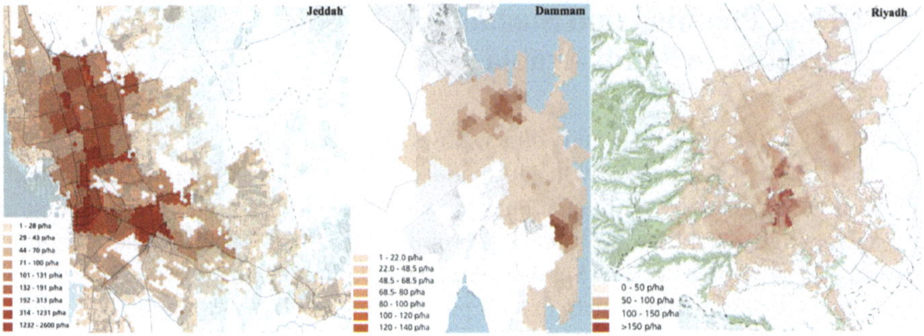

Fig. 4.3 Current distribution of population density in Jeddah, Dammam and Riyadh. *Source* UN-Habitat (2019b, 2019c, 2019d)

alarming demographic increase in urban areas has intensified issues such as urban deprivation, which has been exacerbated by unemployment, underemployment, diminishing real incomes due to persistent inflation, and unregulated migration (Alshebli, 2018; UN-Habitat, 2022). Rapid and unchecked urban expansion poses a significant threat to the sustainable planning of cities in Saudi Arabia, causing a range of socio-economic and environmental challenges (Aina et al., 2019; Ministry of Municipality & Rural Affairs, 2016). The subsequent sections analyze urban planning and central sustainability challenges, and how inclusivity can be achieved via urban planning.

4.4.1 Overview of the Saudi Urban Planning Context

Urban planning in Saudi Arabia is highly centralized, historically dictated by its modernization agenda since 1932, which spurred rapid urbanization and economic growth especially in major cities of Riyadh, Jeddah, and Dammam (Garba, 2004). Decision-making authority over urban planning and management remained highly concentrated at the national level since the twentieth century. While municipal authorities existed, their powers were quite restricted and subordinate to central agencies (Mandeli, 2008; Mubarak, 2004). Major policies influencing urban growth like land grants and loans for housing are directed from the central government, contributing to urban sprawl (Al-Hathloul & Mughal, 2004). Figure 4.4 shows the largely uncontrolled urban expansion across the three case study cities due to urban planning inadequacies. As a result, local development needs and issues are inadequately addressed through decisions that are made exclusively at the national level (Aina et al., 2019). The centralized system creates unnecessary bureaucracy that limits prompt responses as all decisions must pass through multiple layers of upper management (Aina et al., 2019).

It is unquestionable that urban planning as a growth management tool and professional field of practice has not realized its full potential to address the complications

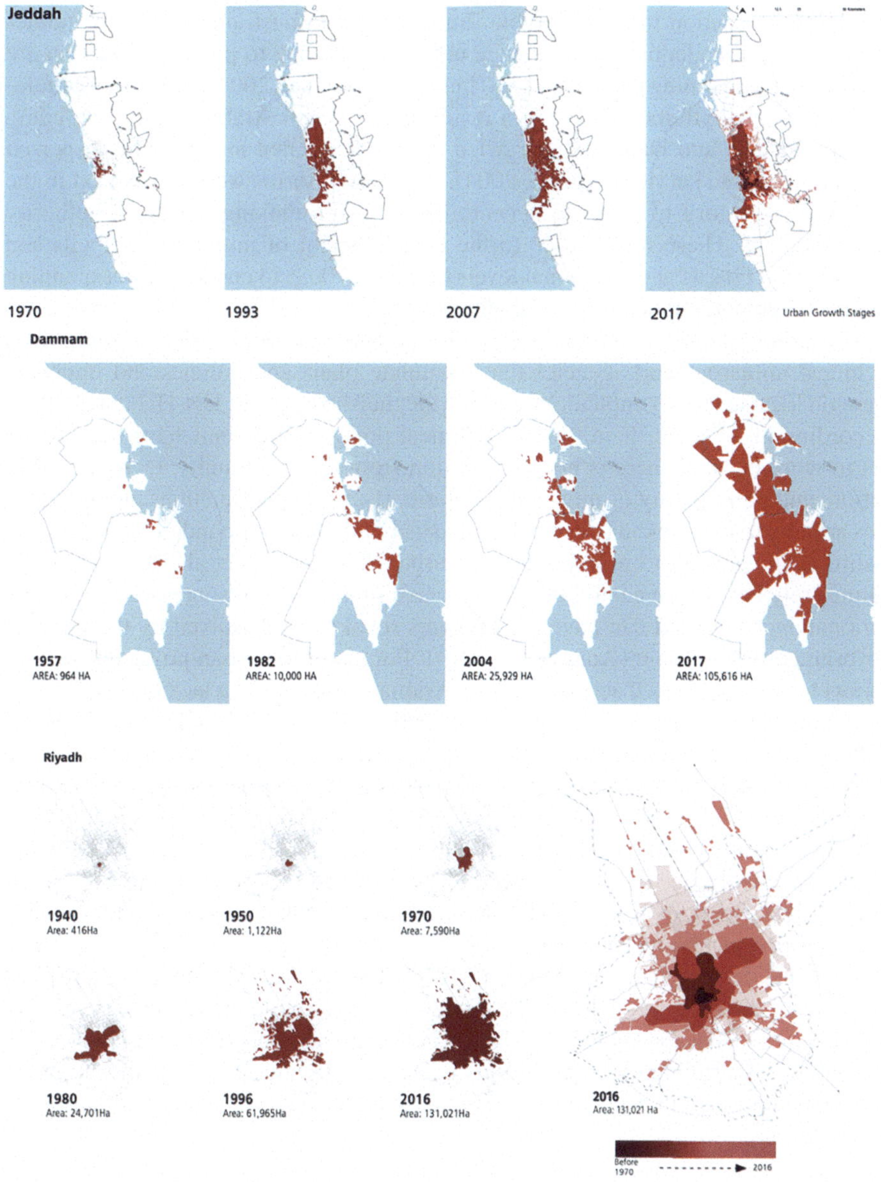

Fig. 4.4 Urban growth stages for Jeddah, Dammam, and Riyadh *Source* Authors' compilation from UN-Habitat (2019a, 2019b, 2019c, 2019d)

of rapid urbanization in Saudi Arabia. Improper planning frameworks and municipal institution inadequacies to manage urban growth due to poor coordination are blamed for the planning incapacity (Al-Hathloul & Mughal, 2004; Ministry of Municipality & Rural Affairs, 2007). As a result, major Saudi Arabian cities, including Riyadh, Jeddah, and Dammam, sprawled in an unregulated manner with dispersed infrastructure and services (Garba, 2004). Planning reforms were introduced in the twenty-first century to promote decentralize decision-making and local authority empowerment. These reforms led to the establishment of municipal councils and development boards at the regional level (Mandeli, 2008) to ensure greater planning and management autonomy of local authorities.

Yet, urban planning and management continue to be dominated by centralized technical ministries and agencies that formulate plans and policies and imposing them on local actors to implement them (Aldegheishem, 2023; UN-Habitat, 2019a). According to Aldegheishem (2023), technical policy instruments have been found not to work because there is no participation from the community, lack of coordination, and poor quality of information. Similarly, planning functions among agencies are overlapping and unclear (Aldegheishem, 2023), compounded by inadequate institutional capacity to respond to rapid urbanization (Aina et al., 2019). Municipalities rely heavily on central budgets, receive funding and resources directly from national agencies, and can have their powers revoked or dissolved by the national government at any time (Aina et al., 2019). For example, urban problems such as sprawl have accumulated across all Saudi Arabian cities due to a lack of coordinated, regionally sensitive policymaking (Garba, 2004; Ministry of Municipality & Rural Affairs, 2007), rendering local authorities powerless in terms of planning action. In Riyadh, our analysis shows a sprawling pattern was observed to represent most of the built-up area, while similar pattern occurred in Jeddah and Dammam extending from the core area toward the northern part and producing disperse pattern of development respectively (see Fig. 4.4). Expectedly, the decentralization of planning systems through these reforms has been characterized with limited success in empowering local governance in practice (Bakhit, 2024).

Genuine decentralization providing local authorities financial autonomy and meaningful decision powers is still a distant reality as national agencies dominates the planning landscape (Mandeli, 2008). The foregoing context has implications for the pursuit of sustainable urbanization. The ensuing sections analyze positionality of urbanization within the context of sustainability and urban planning limitations in Saudi Arabia.

4.4.2 Central Sustainability Inadequacies in Saudi Arabia's Urbanization Management

Our analysis shows that sustainability inadequacies relating to urbanization are similar across Saudi Arabian cities. As a result, the analysis presented within this

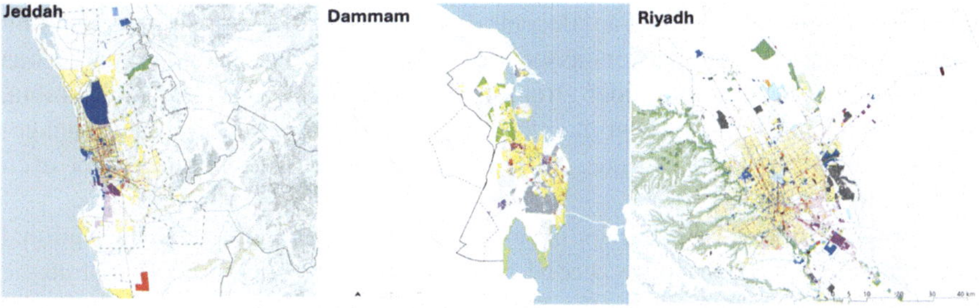

Fig. 4.5 Existing land use for Jeddah, Dammam, and Riyadh *Source* Authors' compilation from UN-Habitat (2019b, 2019c, 2019d)

chapter focuses more on Saudi Arabian cities and less on the three case study cities on how population increase causes urban sprawl, an important area of attention when looking into the difficulties of urban sustainability in Saudi Arabia. In lieu of the rise of the urban population, an unprecedented rate of land resource exploitation has grown substantially in line with the growing urban development, comprising both core and peripheral areas (Alqurashi, 2021). The current land-use maps of the three case study cities show a more rapid consumption of urban centers and surrounding regions as the physical advancement of metropolitan areas witnessed unprecedented expansion (see Fig. 4.5) which is consistent with other Saudi Arabian cities (Ministry of Municipality & Rural Affairs, 2016; UN-Habitat, 2019a).

Inadequate administrative capacity or processes, particularly imprecise land valuation, hinder the ability to make sound zoning choices, which in turn leads to irregular patterns of density, unchecked growth, and informal development as demonstrated in the three case studies (Figs. 4.3 and 4.4). Large-scale urban sprawl, inefficient use of natural resources, and rising infrastructure and service demand are consequently typical in cities. For example, leapfrog development and the growth of dispersed settlements on the periphery of all the three case study cities are contributing to a relative diminution of the significance of the central residential districts. According to Aina and colleagues (2008) and UN-Habitat (2019a), in the last century, Riyadh has grown from a traditional city that was only one square kilometer in size to a metropolis that is three thousand square kilometers in size. This growth has occurred at a pace that is unparalleled anywhere else on Earth. Similarly, the urban area of Jeddah increased dramatically from 18,315 to roughly 54,175 hectares between 1964 and 2007. Others (e.g., Abou-Korin, 2011; Abou-Korin & Al-Shihri, 2015; Alhowaish, 2015; UN-Habitat, 2019a) have reported that the built-up area of Dammam increased from 2200 hectares in 1972 to 25,000 hectares in 2004, expanding by 11 times in just three decades. In comparison, the population grew by only 4.6 times over that same period. Urban sprawl is a problem in Saudi Arabia even for medium- and small-sized cities. With fewer than 400,000 residents, Ha'il city, for instance, saw growth of more than 3.5 times in 2019 compared to 2004 (UN-Habitat, 2019a).

Evaluating the fundamental governance frameworks and land markets that possess substantial implications on land-use patterns and urban growth is crucial.

Although multiple attempts toward decentralization are made, further understanding is achieved regarding the barriers associated with governance and land markets as this reveals a look into the wider structure of local governance in Saudi Arabia, whereby the central government maintains an essential function in urban management and planning as it continues to hold the position of dominant power even regarding local matters.

Shaping the future of urban planning and governance in Saudi Arabia the National Development Plan has been succeeded by Saudi Arabia's Vision 2030, a strategic document that aims to leverage the Saudi Arabia's key status in the Islamic and Arab worlds and position itself as a global economic force. By 2030, Saudi Vision 2030 aspires to transform the nation into an economically successful entity with a prosperous administration. Achieving this vision will require significant changes in goals and viewpoints, particularly in the areas of urban planning and governance. For instance, Saudi Vision 2030 presents goals to address waste management; however, admits that waste management enormously affects the health and wellness of the people and the environment, thus rendering it indispensable to develop frameworks and policies for improvement. Defective planning systems are made worse by waste management. The volume of municipal solid trash production has risen dramatically due to an upsurge in urbanization and higher living standards. With 33 million Saudi Arabian residents, roughly 15 million metric tons of untreated trash are produced yearly, mostly handled by municipal sewage systems or septic tanks (Elboughdiri et al., 2022). Approximately 6 million metric tons of trash are generated yearly from major cities including Riyadh, Dammam, and Jeddah (Elboughdiri et al., 2022). As such, one of the goals of Vision 2030 is reducing pollution-causing waste and putting forth recycling initiatives that would be recognized and practiced at large. However, Zeng et al. (2022) and Abdulghaffar and Williams (2021) recognize the challenge behind these efforts since suboptimal urban management, insufficient infrastructure, and constraints concerning resources still persist.

4.4.3 Sustainable Urbanization: The Challenge of Inclusivity in Saudi Arabia

Given the highly centralized nature of urban planning in Saudi Arabia, our analysis shows that inclusivity is a major barrier to sustainable urbanization across cities beyond the three case cities. A primary concern of inclusivity relates to the provision of basic urban services and infrastructure by the government, a need that has become increasingly pressing amidst rapid urbanization. Saudi Arabia's low population density poses a challenge for the government in delivering basic amenities to widely scattered residential areas. Low-density development as demonstrated in three case study cities is fueled by sprawling horizontal growth (see Fig. 4.3). Many homes are not connected to roads and lack access to essential services such as water, healthcare, and educational resources (Aljoufie, 2021; UN-Habitat, 2022).

Our analysis identifies numerous inefficiencies relating to increased energy and transportation costs, increased expenditure on public space and infrastructure upkeep, reduced benefits of urban clusters, and impeded city productivity. For example, polarized development in Damman is producing inequalities through socio-economic segregations, including development of private compounds and gated communities. While substantial infrastructure developments have improved essential services generally, local governments often struggle with limited resources, lack of state direction, or insufficient power to address these challenges via urban planning has limited options for inclusivity. Effective local governance requires a hands-on approach in planning, budget allocation, management, and overseeing service provision (UN-Habitat, 2016).

On transportation, Dano and Alqahtany (2019) note that Saudi Arabia has 219 cars per 1000 inhabitants, reflecting a culture in which private vehicle use is deeply ingrained. In Riyadh, polarized service provision has limited diversity opportunities and increased reliance on private automobiles, contributing to environmental pollution. This has been realized through land-use strategies and urban designs that allocate substantial space for roads and roadside parking, coupled with the growth of car-centric suburbs. In addition to traffic accidents, dense traffic in metropolitan areas is linked to local health concerns and air pollution (Alotaibi & Potoglou, 2018).

Similarly, Saudi Arabia's housing crisis is driven predominantly by a lack of affordable housing options. While high-rent accommodations are plentiful, affordable housing remains scarce for low-income families, underscoring the need for housing solutions that are both economically accessible and aligned with Saudi Vision 2030. In recent years, the government significantly increased the Real Estate Development Fund's annual allocation from US$13.5 billion in 2008 to US$62.1 billion in 2013 (Ministry of Municipal, Rural Affairs & Housing, 2020; UN-Habitat, 2019a). Yet, as illustrated in Fig. 4.6, housing shortages persist amidst high housing demand. It is anticipated that 1250 million new homes and repairs to subpar homes are required (Ramady, 2021; UN-Habitat, 2019a). Although more recent zoning laws have increased residential neighborhood densities to four stories, they are still low by worldwide or national measures and raise the cost of infrastructure for housing construction (Ramady, 2021; UN-Habitat, 2019a). With approximately 80% of unmet housing needs coming from low- and middle-income families, many housing units remain vacant due to the gap between asking prices and household incomes (UN-Habitat, 2019a).

4.4.4 The Threat of Environmental Problems to Sustainable Urbanization in Saudi Arabia

In this section, analysis of the environmental challenges induced by rapid urbanization and climate confronting Saudi Arabian cities is presented. First, inadequate water is an urgent issue in Saudi Arabia due to limited rainfall, extensive consumption of

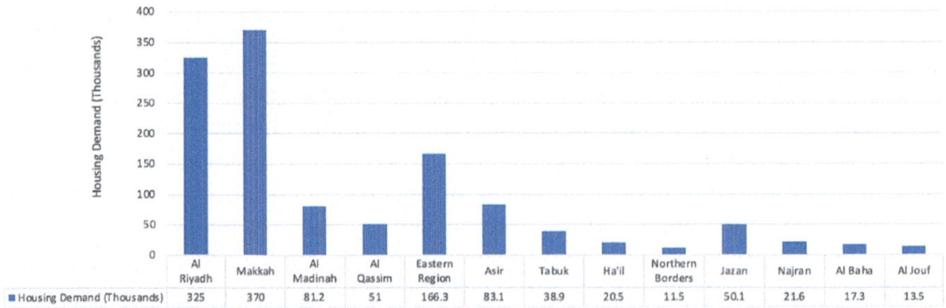

Fig. 4.6 Housing demand in the Saudi Arabia. *Source* Authors' compilation from UN-Habitat (2019a)

subterranean aquifers (Almulhim & Abubakar, 2023; World Bank Climate Change Knowledge Portal, 2021), increasing evaporation, and even decline in freshwater reserves (Odhiambo, 2017). Across the three case study cities, Damman is faced with the lowest freshwater resource endowment not only in Saudi Arabia but globally despite having a significant part of its water resources from fossil groundwater in sedimentary aquifers (Almulhim & Abubakar, 2024; UN-Habitat, 2016). Extreme sandstorms and heat events also affect freshwater access in Riyadh. The pressures of urbanization are further straining the nation's water supply. With a daily capacity of 5 million cubic meters, Saudi Arabia is the world leader in both the production and consumption of desalinated water (Hameed, 2018). As part of Vision 2030 and NTP 2020, the Saudi Arabia aims to encourage the best possible use of the nation's water resources and to enhance service quality by keeping an eye on water pipeline infrastructure to stop leaks. By 2020, the nation hopes to cut distribution-related water loss by 10%, from 25 to 15% (UN-Habitat, 2016).

Similarly, urban heat islands persist and continue to increase across the country due to heat produced by human activity, large buildings, and porous surfaces. For example, the rife nature of urban heat island effect in Damman is attributed to the discontinuity of green networks and the general greenspace deficiency as well as climate change (UN-Habitat, 2019a). This research further expands on its analysis based on reports from Al-Maamary et al. (2017) that Saudi Arabia's average temperature has climbed considerably, about 4 °C throughout the past 40 years, and the Saudi Arabia's coastal regions are anticipated to suffer an extra 2–2.75 °C spike. Cities tend to have summer temperatures reaching as much as 52 °C mainly because of human activity, motor vehicle pollution, and lack of greenery (Abubakar & Aina, 2018). Climate change-induced temperature increases, and prolonged droughts exacerbate this issue by intensifying the strain on limited water resources (World Bank Climate Change Knowledge Portal, 2021). Moreover, urbanization renders 94% of buildings susceptible to coastal flooding globally (Mason et al., 2020). With climate change causing sea levels to rise, low-lying coastal areas in Saudi Arabia, which boasts a 2640-km coastline, face increased risks of urban flooding. Notably, coastal cities such as Dammam and Jeddah have experienced significant flood events, with the

2009 Jeddah flood being the most severe, devasting and impactful in three decades (UN-Habitat, 2019a).

The National Spatial Strategy, developed in 2001 under the direction of the Ministry of Municipal and Rural Affairs, strongly emphasizes the necessity of preserving the environment, reducing urban sprawl, and encouraging spatially balanced growth. The National Spatial Strategy provides strategic alignment programs and initiatives to help transform and expedite the country's climate change mitigation efforts. The recommended formal regulatory framework calls for, but is not confined to, implementing sustainable economic ethics, altering the Saudi Arabia's energy mix to favor clean energy and nuclear power sources, managing global climate risks by boosting the adaptability of urban habitation and activities, and halting land degradation through testing cutting-edge technologies such as cloud ionizing radiation and energy panels for desert agriculture (Gorjian et al., 2022; UN-Habitat, 2019a). But the current disaster management method in Saudi Arabia is primarily reactive instead of proactive, as stated by Alamri (2010). Many people and communities have voiced their disapproval of the Saudi Arabia about the rules and processes utilized to prepare for flash floods. Emergency planning needs in the Saudi Arabia must be investigated, and the obstacles preventing disaster management from working efficiently removed (Alrehaili, 2021).

For Saudi Arabia to have resilient cities and sustainable urbanization, there needs to be an ongoing discourse about how to strengthen essential areas, including land-use planning, infrastructures, governance, and community involvement (Almulhim & Cobbinah, 2024b; Alrehaili, 2021; Ledraa & Al-Ghamdi, 2020). According to Ledraa and Al-Ghamdi (2020), one of the biggest problems is disaster management in the Saudi context which focuses primarily on recovery and relief activities rather than prevention or mitigation. They also argue that the present disaster management method is less concerned with management and more concerned with catastrophe containment. These claims are emphasized by Alrehaili (2021) who asserts that ambiguous legislation and plans, inadequate coordination among stakeholders, and a dearth of databases hinder disaster management in the Saudi Arabia.

4.5 Discussion: Toward Sustainable Urbanization: The Centrality of Planning

In Saudi Arabia, urbanization has served as a catalyst for growth reflected in economic opportunities and infrastructural enhancements. However, urbanization has created significant urban challenges as our analysis shows that major urban agglomerations, particularly Riyadh, Jeddah, and Dammam, are experiencing increased negative externalities, including inequality, environmental deterioration, unaffordable housing, water scarcity, and uncontrolled sprawl. Such impacts were connected, to a greater degree, to the inadequate purposive urban planning-guided agenda reflected in limited participation, inadequate awareness and communication, and inclusivity limitations. Such urban planning inadequacy conditions has limited Saudi Arabia's capacity to effectively organize and respond to the complications of urbanization, particularly for its metropolitan cities.

Moreover, inclusion issues were made worse by political officials' ineffective communication with the urban community, specifically regarding ensuring fundamental services. As such, a lack of public participation occurs, resulting in uncontrolled sprawling of cities and associated low-density development as well as challenges with infrastructure. For instance, cases of lack of housing availability and infrastructure support and management across Riyadh, Jeddah, and Dammam have exacerbated these issues as the number of people moving into metropolitan continues to grow. This further expands other issues, including overcrowding, congestion, and greater strain on the existing infrastructure and housing, the natural environment, and even new infrastructures, which affect the ecosystem (Trask, 2022). This creates an ecological imbalance with adverse environmental impacts, including the heightened possibility of risks and disasters (Almulhim & Cobbinah, 2023b; Cobbinah et al., 2015; Mersal, 2016). Thus, there was a disconnect between government efforts and urban community positionality in relation to managing urbanization complications, particularly relating to climate change in an uncontrolled sprawl environment. The findings of this chapter lend credence to those reported by Maizland (2021) in China, highlighting the compounding nature of greenhouse gas emissions by urbanization and associated health risks posed by pollution, soil contamination, and water scarcity.

Our analysis supports those of previous studies. For example, Basu and Das (2021) found that environmental management in global south countries such as India is still not given substantial consideration in terms of the management and development of urban locations amidst air pollution, water pollution, and improper waste disposal struggles. Similarly, Aliyu and Amadu (2017) found that Nigeria's urbanization is demographically driven, insufficient in socio-economic dividends or improvements to the urban environment. This has led to urban health crises such as unsanitary conditions and shanty settlements, insufficient solid waste management, an inadequate safe water supply, a doubled prevalence of disease, and congested, inefficient, and dangerous transport systems, all of which suffer from inadequate government action.

Our analysis further reveals that a lack of coordination among planning authorities contributes to a planning process that moves at a glacial pace compared to the rapid expansion of cities. As noted in the three case study cities, the cumulative stress on ecosystems associated with urbanization, industrialization, water scarcity, and climate change is substantial, creating a pressing need for measures that could stop or drastically cut marine pollution such as waste and chemicals. This is the case when prevention of coastal infiltration necessitates the integration of methods to boost socio-economic growth, food security, and the mitigation of hazards from industrial emissions. This includes ecosystem-based resilience response, which entails such activities as managing, conserving, and restoring ecosystems in a sustainable manner so that they can continue to supply services that aid humans in coping with the negative impacts of climate change. Additionally, our analysis reveals that political centralization in urban planning complicates and hinders sustainability efforts.

Further access to accurate, ongoing, and timely data on nationwide trends in urban planning and city development across time and space is essential for urban and regional policymakers and planners to thoroughly understand the dynamics of urbanization, to anticipate the impacts of rapid urban growth, and make more precise forecasts concerning sustainable urban development (Li & Zhou, 2017). Saudi Arabia's economy and society depend significantly on its ecological resources for energy production, food cultivation, and raw materials for industrial processes, but environmental degradation is set to intensify unless remedial actions are taken. In summary, the study reveals that urbanization is a complex sustainable development challenge, not a minor urban issue. Managing urbanization in Saudi Arabian cities necessitates a comprehensive approach that prioritizes urban planning that responds to the welfare of vulnerable urban residents and the future of cities.

4.6 Conclusion

The case studies of Jeddah, Dammam, and Riyadh demonstrate how important urban planning is in tackling the unique problems brought on by Saudi Arabia's fast urbanization. Our findings show that although tremendous attempts have been made to direct urban growth through comprehensive planning, these initiatives have frequently been thwarted by the preference for economic growth over sustainable practices, centralized control, and insufficient local implementation.

The urban planning literature has called for a better understanding of planning-led development and its contribution to sustainable urbanization (Cobbinah & Gaisie, 2023; Mersal, 2016; Tibaijuka, 2006). Key strategies identified in the case studies include boosting public engagement in planning processes, decentralizing decision-making to empower local authorities, and integrating environmental sustainability into urban development plans. These measures are vital for minimizing urban sprawl, lowering socio-economic inequities, and managing environmental concerns such as water scarcity and pollution. A need to engage in city-level examination of the efforts of governments to achieve sustainable urbanization is necessary to move beyond romantic views of problematizing urbanization to a planning-guided response. Such research needs to offer a better insight of the urbanization dynamics and mechanisms via which urban planning can positively contribute to sustainable urbanization.

The study shows that a change from a top-down to a more inclusive and locally responsive approach is necessary for Saudi Arabia to achieve sustainable urbanization. This entails encouraging cooperation between the national and local governments, involving the public in the decision-making process, and making sure that urban planning adheres to sustainable development principles.

In this chapter, a sustainable development approach was taken to understand urbanization processes in Saudi Arabia. Such an approach suggests that for the attainment of sustainable urbanization, government efforts require urban planning-guided elements that are conducive to promoting inclusivity, addressing urban sprawl, and managing environmental challenges. This means that planning-led urban planning should not just happen to address challenges of urbanization but rather should be based on the foundational motivation of sustainable development that seeks the overall urban good.

The relevance of urban planning-guided approach to urbanization management lies in its capacity to explain how the urban space is predisposed to the threat of unplanned urbanization and at the same time offers a purposive interest and opportunities for addressing the problems of urbanization. The purposive interest of urban planning-led approach reflects both the dangers and opportunities that urbanization presents and how they impact the sustainability of the urban space.

Cobbinah et al., (2015 p. 69) noted:

> the nature of urbanization has, to a limited extent, some positive implications for socio-economic development and management …, including sustainable social, economic and environmental transformation. Unfortunately, the negative implications of urbanization dominate and abound … The limited positive implications of urbanization have been overshadowed by unsustainable land development, emergence of informal settlements (e.g., slums), and increased urban poverty characterised by unemployment, dominance of informal economic activities (e.g., hawking), vulnerability and insecurity, and inadequate water supply and sanitation.

This chapter contributes to such understanding from a sustainable development perspective by noticing desired elements and processes for achieving sustainable urbanization, but urban planning inadequacies have stifled progress. These urban planning inadequacies contribute to compounding urbanization problems, particularly sprawl, despite apparent effort by governments to address the urbanization challenge as demonstrated through the Saudi Arabian situation.

The acknowledgment and engagement of purposive urban planning-guided approach in global south countries, including Saudi Arabia, is possible through political diagnosis. This can be achieved through mechanisms examining the tripartite foci of sustainable development including levels of participation and scales of involvement, social inclusion, and improved levels of awareness and communication. Similarly, national and local governments can assess how these urban planning-guided mechanisms are advancing sustainable urbanization by further identifying and examining existing formal and informal associations in the urban space and their motivations and aspirations toward their cities. In the case of sustainable urbanization, such efforts can be traced toward joint efforts between governments and the urban (metropolitan) community resulting in urbanization planning and management strategies aimed at improving urban functionality.

This chapter identified some urban planning-guided elements that could spur sustainable urbanization. Future research should examine the mechanisms promoting and achieving polycentric urban design to encourage increasing densities at multiple urban centers rather than concentrating densities only in the larger cities in an uncontrolled sprawled manner. Further, the role of urban residents in promoting inclusivity ideas on basic service provision that lead to sustainable urbanization should be researched. Studies should be aimed at understanding the extent of sustainable development principles adoption or rejection of in both urbanization management initiatives and implementation as well as their outcomes. Finally, research should examine the role of both local and national governments in their engagement with purposive urban planning-guided practices leading to sustainable urbanization.

The research findings and emerging knowledge from further studies should inform governments, practitioners, researchers, and urban communities interested in seeking urban planning-guided mechanisms to achieving sustainable urbanization. The application of this information allows a deeper understanding of the sustainable urbanization dynamics reflecting the interests and aspirations of the urban community and the government. Attempts should be made by researchers, governments, and practitioners to improve their understanding of sustainable urbanization processes intrinsic to urban planning in cities and regions faced with rapid urbanization. Without such understanding, even the well-intentioned efforts could likely be hindered by impetuously attempting to implement strategies targeted at sustainable urbanization in the global south.

Acknowledgements This chapter is based on the following previously published article and is reproduced with permission from Elsevier: Almulhim, A. I., & Cobbinah, P. B. (2024). Urbanization provocateur: Reaching urban planning-led development in Saudi Arabia. *Land Use Policy, 147*, 107365.

References

Abdulghaffar, N. A., & Williams, I. D. (2021). Development of sustainable waste management in higher education institutions. *AIMS Environmental Science, 8*(3), 238–254. https://doi.org/10.3934/environsci.2021016

Abou-Korin, A. A., & Al-Shihri, F. S. (2015). Rapid urbanization and sustainability in Saudi Arabia: The case of Dammam metropolitan area. *Journal of Sustainable Development, 8*(9), 52. https://doi.org/10.5539/jsd.v8n9p52

Abou-Korin, A. A. (2011, December). Impacts of rapid urbanisation in the Arab World: The case of Dammam metropolitan area, Saudi Arabia. In *5th Int'l Conference and Workshop on Built Environment in Developing Countries (ICBEDC 2011)* (pp. 1–25). University Sains Malaysia.

Abubakar, I. R., & Aina, Y. A. (2018). Achieving sustainable cities in Saudi Arabia: Juggling the competing urbanization challenges. In *E-planning and collaboration: Concepts, methodologies, tools, and applications* (pp. 234–255). IGI Global.

Abubakar, I. R., & Dano, U. L. (2020). Sustainable urban planning strategies for mitigating climate change in Saudi Arabia. *Environment, Development and Sustainability, 22*(6), 5129–5152. https://doi.org/10.1007/s10668-019-00417-1

Aina, Y. A., Wafer, A., Ahmed, F., & Alshuwaikhat, H. M. (2019). Top-down sustainable urban development? Urban governance transformation in Saudi Arabia. *Cities, 90*, 272–281. https://doi.org/10.1016/j.cities.2019.03.003

Al Bassam, A. M. A. (2012). *Urbanisation and migration in Saudi Arabia: The case of Buraydah City* [Unpublished Doctoral Dissertation]. University of Leicester, Leicester.

Alamri, Y. A. (2010). Emergency management in Saudi Arabia: Past, present and future. *Un. Of Christchurch report, New Zealand, 21.*

Aldegheishem, A. (2023). Urban growth management in Riyadh, Saudi Arabia: An assessment of technical policy instruments and institutional practices. *Sustainability, 15*(13), 10616.

Alharthi, B., & El-Damaty, T. A. (2022). Study the urban expansion of Taif City using remote sensing and GIS techniques for decision support system. *Advances in Remote Sensing, 11*(1), 1–15.

Al-Hathloul, S. A., & Anis-ur-Rahmaan, S. (1985). The evolution of urban and regional planning in Saudi Arabia. *Ekistics, 52*, 206–212.

Al-Hathloul, S., & Mughal, M. A. (2004). Urban growth management—The Saudi experience. *Habitat International, 28*(4), 609–623.

Alhowaish, A. K. (2015). Eighty years of urban growth and socioeconomic trends in Dammam Metropolitan Area, Saudi Arabia. *Habitat International, 50*, 90–98.

Aliyu, A. A., & Amadu, L. (2017). Urbanization, cities, and health: The challenges to Nigeria—A review. *Annals of African Medicine, 16*(4), 149.

Aljoufie, M. (2021). The impact assessment of increasing population density on Jeddah road transportation using spatial-temporal analysis. *Sustainability, 13*(3), 1455. https://doi.org/10.3390/su13031455

Al-Maamary, H. M. S., Kazem, H. A., & Chaichan, M. T. (2017). Climate change: The game changer in the Gulf Cooperation Council Region. *Renewable and Sustainable Energy Reviews, 76*, 555–576. https://doi.org/10.1016/j.rser.2017.03.048

Almulhim, A. I., & Abubakar, I. R. (2023). Developing a sustainable water conservation strategy for Saudi Arabian cities. *Groundwater for Sustainable Development, 23*, 101040.

Almulhim, A. I., & Abubakar, I. R. (2024). A segmentation approach to understanding water consumption behavioral patterns among households in Saudi Arabia for a sustainable future. *Resources, Environment and Sustainability, 15*, 100144. https://doi.org/10.1016/j.resenv.2023.100144

Almulhim, A. I., Sharifi, A., Aina, Y. A., Ahmad, S., Mora, L., Filho, W. L., & Abubakar, I. R. (2024). Charting sustainable urban development through a systematic review of SDG11 research. *Nature Cities*. https://doi.org/10.1038/s44284-024-00117-6

Almulhim, A. I., & Cobbinah, P. B. (2023a). Can rapid urbanization be sustainable? The case of Saudi Arabian cities. *Habitat International, 139*, 102884. https://doi.org/10.1016/j.habitatint.2023.102884

Almulhim, A. I., & Cobbinah, P. B. (2023b). Urbanization-environment conundrum: An invitation to sustainable development in Saudi Arabian cities. *International Journal of Sustainable Development & World Ecology*. https://doi.org/10.1080/13504509.2022.2152199

Almulhim, A. I., & Cobbinah, P. B. (2024a). Ontological pluralism in Saudi Arabia's sustainable urbanization framing. *Habitat International, 151*, 103139. https://doi.org/10.1016/j.habitatint.2024.103139

Almulhim, A. I., & Cobbinah, P. B. (2024b). Framing resilience in Saudi Arabian cities: On climate change and urban policy. *Sustainable Cities and Society, 101*, 105172. https://doi.org/10.1016/j.scs.2024.105172

Alotaibi, O., & Potoglou, D. (2018). Introducing public transport and relevant strategies in Riyadh City, Saudi Arabia: A stakeholders' perspective. *Urban, Planning and Transport Research, 6*(1), 35–53. https://doi.org/10.1080/21650020.2018.1463867

Alqurashi, A. F. (2021). Quantification of urban patterns and processes through space and time using remote sensing data: A comparative study between three Saudi Arabian cities. *Sustainability, 13*(22), 12615. https://doi.org/10.3390/su132212615

Alrehaili, N. (2021). An investigation into emergency planning requirements and challenges of disaster management in the Kingdom of Saudi Arabia. *International Journal of Disaster Management, 4*(3), 1–10.

Alshebli, A. (2018). *Towards making urban planning practices more effective amid rapid urban growth in Riyadh-Saudi Arabia* (Doctoral dissertation). University of Birmingham.

Archistar. (2024). The importance of urban planning—The seven key reasons. Arhistar Pty Ltd. Retrieved from https://www.archistar.ai/blog/the-importance-of-urban-planning-the-seven-key-reasons/#popup-id-31813. Accessed 16 May 2023.

asghar Pilehvar, A. (2021). Spatial-geographical analysis of urbanization in Iran. *Humanities and Social Sciences Communications, 8*(1), 1–12. https://doi.org/10.1057/s41599-021-00741-w

Bai, X., McPhearson, T., Cleugh, H., Nagendra, H., Tong, X., Zhu, T., & Zhu, Y. G. (2017). Linking urbanization and the environment: Conceptual and empirical advances. *Annual Review of Environment and Resources, 42*(1), 215–240. https://doi.org/10.1146/annurev-environ-102016-061128

Bakhit, W. I. (2024). *Urban planning trends and challenges in the pursuit of Saudi Vision 2030.* King Salman Center for Local Governance. http://www.ksclg.org/en/publication-project/urban-planning-trends-and-challenges-in-the-pursuit-of-saudi-vision-2030/. Accessed 16 May 2023.

Basu, T., & Das, A. (2021). Systematic review of how eco-environmental transformation due to urbanization can be investigated in the sustainable development of Indian cities. *Environmental Challenges, 4*, 100099.

Black, D., & Henderson, V. (1999). A theory of urban growth. *Journal of Political Economy, 107*(2), 252–284.

Central Intelligence Agency. (2019). *Saudi Arabia.* https://www.cia.gov/the-world-factbook/countries/saudi-arabia/ https://www.cia.gov/the-world-factbook/countries/saudi-arabia/. Accessed 16 May 2023.

Chen, M., Zhang, H., Liu, W., & Zhang, W. (2014). The global pattern of urbanization and economic growth: Evidence from the last three decades. *PLoS ONE, 9*(8), e103799.

Cobbinah, P. B. (2017). Managing cities and resolving conflicts: Local people's attitudes towards urban planning in Kumasi, Ghana. *Land Use Policy, 68*, 222–231.

Cobbinah, P. B., Asibey, M. O., Opoku-Gyamfi, M., & Peprah, C. (2019). Urban planning and climate change in Ghana. *Journal of Urban Management, 8*(2), 261–271.

Cobbinah, P. B., & Darkwah, R. M. (2017). Urban planning and politics in Ghana. *GeoJournal, 82*, 1229–1245.

Cobbinah, P. B., Erdiaw-Kwasie, M. O., & Amoateng, P. (2015). Africa's urbanisation: Implications for sustainable development. *Cities, 47*, 62–72.

Cobbinah, P. B., & Gaisie, E. (Eds.). (2023). *Reimagining urban planning in Africa.* Cambridge University Press.

Dano, U. L., & Alqahtany, A. M. (2019). Issues undermining public transport utilization in Dammam city, Saudi Arabia: An expert-based analysis. *Journal of Sustainability Science and Management, 14*(2), 155–169.

DeNicola, E., Aburizaiza, O. S., Siddique, A., Khwaja, H., & Carpenter, D. O. (2015). Climate change and water scarcity: The case of Saudi Arabia. *Annals of Global Health, 81*(3), 342–353. https://doi.org/10.1016/j.aogh.2015.08.005

Elboughdiri, N., Khan, M. I., & Missaoui, T. (2022). Estimating the energy recovery potential from municipal solid waste in Ha'il Region (Kingdom of Saudi Arabia). *Yanbu Journal of Engineering and Science, 19*(1), 11–17. https://doi.org/10.53370/001c.31893

Elmqvist, T., Andersson, E., McPhearson, T., Bai, X., Bettencourt, L., Brondizio, E., Colding, J., Daily, G., Folke, C., Grimm, N., Haase, D., Parnell, S., Polasky, S., Seto, K. C., & van der Leeuw, S. (2021). Urbanization in and for the Anthropocene. *Npj Urban Sustainability, 1*(1), 1–6. https://doi.org/10.1038/s42949-021-00018-w

Fuseini, I., & Kemp, J. (2015). A review of spatial planning in Ghana's socio-economic development trajectory: A sustainable development perspective. *Land Use Policy, 47*, 309–320.

Garba, S. B. (2004). Managing urban growth and development in the Riyadh metropolitan area, Saudi Arabia. *Habitat International, 28*(4), 593–608.

Geneva Environment Network. (2022). *Cities and the environment.* https://www.genevaenvironmentnetwork.org/resources/updates/cities-and-the-environment/. Accessed 16 May 2023.

Gorjian, S., Ahmed, M., Fakhraei, O., Eterafi, S., & Jathar, L. D. (2022). Solar desalination technology to supply water for agricultural applications. In *Solar energy advancements in agriculture and food production systems* (pp. 271–311). Academic Press. https://doi.org/10.1016/B978-0-323-89866-9.00002-X

Hameed, S. (2018). Investment in renewable energy in the GCC states: Status and policy options. In *Contemporary Persian Gulf* (pp. 23–37). Routledge.

Kahouli, B., Miled, K., & Aloui, Z. (2022). Do energy consumption, urbanization, and industrialization play a role in environmental degradation in the case of Saudi Arabia? *Energy Strategy Reviews, 40*, 100814.

Kuncic, A., & Andrés, M. (2021). *Assessment of Sustainable Development Goals in Saudi Arabia.* United Nations Saudi Arabia.

Ledraa, T. A., & Al-Ghamdi, A. M. (2020). Planning and management issues and challenges of flash flooding disasters in Saudi Arabia: The case of Riyadh City. *Journal of Architecture and Planning, 32*, 155–171.

Li, X., & Zhou, Y. M. (2017). Offshoring pollution while offshoring production? *Strategic Management Journal, 38*(11), 2310–2329.

Maizland, L. (2021). *China's fight against climate change and environmental degradation.* Council on Foreign Relations. https://www.cfr.org/backgrounder/china-climate-change-policies-environmental-degradation. Accessed 16 May 2023.

Malkina-Pykh, I. G. (2002). Integrated assessment models and response function models: Pros and cons for sustainable development indices design. *Ecological Indicators, 2*(1–2), 93–108.

Mandeli, K. N. (2008). The realities of integrating physical planning and local management into urban development: A case study of Jeddah, Saudi Arabia. *Habitat International, 32*(4), 512–533.

Mason, D., Iida, A., Watanabe, S., Jackson, L. P., & Yokohari, M. (2020). How urbanization enhanced exposure to climate risks in the Pacific: A case study in the Republic of Palau. *Environmental Research Letters, 15*(11), 114007. https://doi.org/10.1088/1748-9326/abb9dc

Ministry of Municipality and Rural Affairs. (2016). *The national report for the third UN conference on housing and sustainable urban development (HABITAT III) for the Kingdom of Saudi Arabia.* https://habitat3.org/wp-content/uploads/FULL-REPORT-ENGLISH.pdf. Accessed 16 May 2023.

Ministry of Municipality and Rural Affairs, Saudi Arabia. (2007). *Local and action area plans for cities of Dammam Metropolitan and Governorates of Qatif and Ras-Tanura.* Prepared for Eastern Province Amanah, MOMRA, Kingdom of Saudi Arabia.

Ministry of Municipal, Rural Affairs and Housing. (2020). Gov.sa. Retrieved 14th July 2024, from https://www.my.gov.sa/wps/portal/snp/agencies/agencyDetails/AC174/!ut/p/z0/04_Sj9CPykssy0xPLMnMz0vMAfIjo8zivQIsTAwdDQz9LQwNzQwCnS0tXPwMvYwNDAz0g1Pz9L30o_ArAppiVOTr7JuuH1WQWJKhm5mXlq8f4ehsaG6iX5DtHg4Ayo0duQ!!/. Accessed 16 May 2023.

Mensah, J. (2019). Sustainable development: Meaning, history, principles, pillars, and implications for human action: Literature review. *Cogent Social Sciences, 5*(1), 1653531.

Mersal, A. (2016). Sustainable urban futures: Environmental planning for sustainable urban development. *Procedia Environmental Sciences, 34*, 49–61.

Mtapuri, O., & Myeni, S. L. (2019). Sustainable development goals and the new urban agenda: A South African experience. In S. L. Myeni & A. Okem (Eds.), *The political economy of government subsidised housing in South Africa* (pp. 12–27). Routledge.

Mubarak, F. A. (2004). Urban growth boundary policy and residential suburbanization: Riyadh, Saudi Arabia. *Habitat International, 28*(4), 567–591.

Muhammad, E. (2021, August 14). Pollution due to urbanization. LinkedIn. https://www.linkedin.com/pulse/pollution-due-urbanization-eiman-muhammad. Accessed 16 May 2023.

Mutuku, C. (2018). *Urbanization in developing countries.* GRIN Verlag. https://www.grin.com/document/432475. Accessed 16 May 2023.

Naess, P. (2001). Urban planning and sustainable development. *European Planning Studies, 9*(4), 503–524.

Odhiambo, G. O. (2017). Water scarcity in the Arabian Peninsula and socio-economic implications. *Applied Water Science, 7*(5), 2479–2492. https://doi.org/10.1007/s13201-016-0440-1

Page, M. J., McKenzie, J. E., Bossuyt, P. M., Boutron, I., Hoffmann, T. C., Mulrow, C. D., … Moher, D. (2021). *The PRISMA 2020 statement: an updated guideline for reporting systematic reviews*. https://www.bmj.com/content/372/bmj.n71. Accessed 16 May 2023.

Ramady, M. (2021). The Saudi capital market: Coming of age. In *Financial regulation and liberation* (pp. 43–57). Springer. https://doi.org/10.1007/978-3-030-68267-5

Rasoolimanesh, S. M., Badarulzaman, N., & Jaafar, M. (2012). City development strategies (CDS) and sustainable urbanization in developing world. *Procedia-Social and Behavioral Sciences, 36*, 623–631.

Roser, M. (2023). *What is economic growth? And why is it so important?* Our World in Data. Available at https://ourworldindata.org/what-is-economic-growth?fbclid=IwAR3puXhY9yDugYkoCwyKH_kiY6rfYKMGaABRnzDjFq3s7MIJ8I3aICXHumM. Accessed 16 May 2023.

Roy, M. (2009). Planning for sustainable urbanization in fast growing cities: Mitigation and adaptation issues addressed in Dhaka, Bangladesh. *Habitat International, 33*(3), 276–286.

Schaefer, A., & Crane, A. (2005). Addressing sustainability and consumption. *Journal of Macromarketing, 25*(1), 76–92.

Shaban, A., Kourtit, K., & Nijkamp, P. (2022). Causality between urbanisation and economic growth: Evidence from the Indian states. *Frontiers in Sustainable Cities, 4*, 901346.

Shen, L. Y., Ochoa, J. J., Shah, M. N., & Zhang, X. (2011). The application of urban sustainability indicators—A comparison between various practices. *Habitat International, 35*(1), 17–29.

Solly, A., Berisha, E., & Cotella, G. (2021). Towards sustainable urbanization. Learning from what's out there. *Land, 10*(4), 356.

Tan, Y., Xu, H., & Zhang, X. (2016). Sustainable urbanization in China: A comprehensive literature review. *Cities, 55*, 82–93.

The World Bank. (2020). *Urban development overview*. https://www.worldbank.org/en/topic/urbandevelopment/overview. Accessed 16 May 2023.

Tibaijuka, A. (2006). *The importance of urban planning in urban poverty reduction and sustainable development*. In World planners congress, Vancouver, June 2006.

Trask, B. S. (2022, May 13). *Migration, urbanization, and the family dimension* [Background Paper]. UNDESA.

UNDP. (2022). *Saudi Arabia report (2022–2023)*. https://www.undp.org/saudi-arabia/publications/undp-saudi-arabia-report-2022-2023. Accessed 16 May 2023.

UNDRR. (2018). *Urbanization trends and urban resilience in the Arab region*. https://www.undrr.org/sites/default/files/2021-11/Chapter5-RARArabRegion2021.pdf?startDownload=true. Accessed 16 May 2023.

UN-Habitat. (2016). *Country profile Saudi Arabia*. Available online https://unhabitat.org/sites/default/files/download-manager-files/Country%20Profile%20Saudi%20Arabia%20English.pdf. Accessed 16 May 2023.

UN-Habitat. (2019a). *Saudi Cities Report 2019*. Available online https://unhabitat.org/sites/default/files/2020/05/saudi_city_report.english.pdf. Accessed 16 May 2023.

UN-Habitat. (2019b). *Dammam city profile*. Available online https://unhabitat.org/dammam-city-profile. Accessed 16 May 2023.

UN-Habitat. (2019c). *Jeddah city profile*. Available online https://unhabitat.org/jeddah-city-profile. Accessed 16 May 2023.

UN-Habitat. (2019d). *Riyadh city profile*. Available online https://unhabitat.org/riyadh-city-profile. Accessed 16 May 2023.

UN-Habitat. (2020). *World Cities Report 2020: The value of sustainable urbanization*. https://unhabitat.org/sites/default/files/2020/10/wcr_2020_report.pdf. Accessed 16 May 2023.

UN-Habitat. (2022). *The State of Arab Cities Report 2022*. https://unhabitat.org/the-state-of-arab-cities-report-2022. Accessed 16 May 2023.

UN-Habitat/DFID. (2002). *Sustainable urbanisation: achieving Agenda 21*. United Nations Human Settlement Programme/Department for International Development, Nairobi.

Whitehead, M. (2003). (Re) Analysing the sustainable city: Nature, urbanisation and the regulation of socio-environmental relations in the UK. *Urban Studies, 40*(7), 1183–1206.

World Bank Climate Change Knowledge Portal. (2021). https://climateknowledgeportal.worldbank.org/country/saudi-arabia. Accessed 16 May 2023.

World Health Organization. (2021, October 29). *Urban health*. https://www.who.int/news-room/fact-sheets/detail/urban-health. Accessed 16 May 2023.

Wu, J. (2010). Urban sustainability: An inevitable goal of landscape research. *Landscape Ecology, 25*, 1–4.

Wu, J. (2014). Urban ecology and sustainability: The state-of-the-science and future directions. *Landscape and Urban Planning, 125*, 209–221.

Yin, R. K. (2015). *Qualitative research from start to finish* (2nd ed.). Guilford Publications.

Zeng, X., Yu, Y., Yang, S., Lv, Y., & Sarker, M. N. I. (2022). Urban resilience for urban sustainability: Concepts, dimensions, and perspectives. *Sustainability, 14*(5), 2481. https://doi.org/10.3390/su14052481

Chapter 5
Desiring Resilience in Saudi Arabian Cities: On Climate Change and Urban Policy

5.1 Introduction

Building resilient cities presents a key challenge confronting countries in the Global South (Almulhim & Cobbinah, 2023a; Ledraa & Al-Ghamdi, 2020). At the turn of the twenty-first century, cities in the Global South experienced unprecedented urban growth and management challenges, including flood events, droughts, unemployment, rising sea levels, urban sprawl, urban poverty, heatwaves, and coastal erosion (Abualnaja, 2011; Alrehaili, 2021; Ameen & Mourshed, 2017; Ewing & Synolakis, 2011; Lam et al., 2016; Mead, 2022). The vulnerability of cities in the Global South is expected to increase as urban growth and management challenges become more frequent and intense, with cities in Africa and Asia being the most affected (Cobbinah & Finn, 2022; Dastjerdi et al., 2021; Ochoa et al., 2018; Salehi & Sharifi, 2022; Sharifi, 2020). Across Asia and Africa, many metropolitan areas experience extreme infrastructure inadequacy and unsustainable levels of urban development (Al-Bassam et al., 2014; Alrehaili, 2021; Alizade & Sharifi, 2020; Bautista-Puig et al., 2022; Busayo et al., 2019; Champlin et al., 2023; Ledraa & Al-Ghamdi, 2020). In the face of urban development catastrophes, meeting growing urban infrastructure demands and working toward sustainable development and resilient futures is becoming increasingly challenging (e.g., Sharifi et al., 2021; Wang, 2022; Zhou et al., 2021).

Scientific research tends to explain the increase in urban growth and management vulnerability of Global South cities because of the expansion of urbanized areas induced by rapid population growth and climate change (Almulhim & Cobbinah, 2023b; Cohen, 2006; Mahdaoui & Alibou, 2021). In most cases, climate change is the force that limits and constrains the sustainability and resilience of cities in the Global South by altering and threatening the survival and functionality of cities (UNCC, 2023). Human geography and urban studies have made important progress in developing approaches to analyze and understand the intensity of climate change and rapid urbanization pressures on urban functionality, sustainability, and resilience

A. I. Almulhim and P. B. Cobbinah, *Urbanization Rethought*, Urban Sustainability,
https://doi.org/10.1007/978-981-92-0118-1_5

in cities in the Global South (Boyd & Juhola, 2015; Newman et al., 2009; Ochoa et al., 2018; Sharifi, 2022b). However, these analyses fail to recognize how urban policy in Global South cities shapes the way in which urban resilience strategies develop and unfold in the most vulnerable regions of the world (Asadzadeh et al., 2022; Cobbinah et al., 2011, 2015; Sharifi, 2019, 2020). The problem of inadequate analysis and embeddedness of urban policy in resilience framing in cities of the Global South is that it often leads to atomistic responses and solutions that are likely to perpetuate urban growth and management challenges, and in turn, reproduce dysfunctional urban spaces and unsustainable climate change management.

In this chapter, we draw on the urban resilience literature and use Saudi Arabian cities to demonstrate that urban policy reflections about climate change are imperative for the pursuit of urban resilience. Urban resilience refers to "the ability of an urban system and its constituent socio-ecological and socio-technical networks across temporal and spatial scales to maintain or rapidly return to desired functions in the face of a disturbance, to adapt to change, and to quickly transform systems that limit current or future adaptive capacity" (Meerow et al., 2016, p. 39). In particular, we draw on critical urban studies and human geography literature to foreground a conceptualization of resilience within and beyond urban studies, which is foundational to advancing debates on addressing urban growth and management challenges in cities of the Global South.

Using Saudi Arabia as a case study, this chapter draws on the growing body of academic work on urban resilience to address the following questions: (a) How resilient are Saudi Arabian cities? (b) How can urban policy contribute to developing more resilient cities in Saudi Arabia? We specifically analyzed land use and urban planning, disaster management, and green infrastructural systems to capture how digital technologies that foster climate resilience, governance, and public participation shape urban policies on climate resilience. Our analysis makes three key contribution to urban studies and environmental planning literature. First, it shifts attention away from a siloed focus on either urbanization or climate change, and instead provides a holistic agenda toward addressing the implications of these two phenomena via urban policy. This approach allows an exploration of the role played by the collision of climate change and rapid urbanization in the production of dysfunctional and unsustainable urban spaces while also assessing cities' ability to respond to crises related to climate change and rapid urbanization via resilience building. Second, this chapter demonstrates that climate change risks are not necessarily less in wealthy global south countries. Despite Saudi Arabia's oil wealth, climate change risks are considerable (Fang & Lerner, 2019). Our analysis adds to the debate on climate change by demonstrating that without an urban policy agenda that reflects the aspirations of local communities, the risks of climate change may increase. Third, this chapter demonstrates the potential of managing climate change risks via urban policy using smart innovations, and further shows the limitations of these emerging innovative frameworks. This is an important contribution to the climate change-urban policy dialogue in the global south (see Wang, 2022), and provides a context for reference and consideration by countries in their efforts toward building resilient futures.

We first analyzed urban strategies and approaches to building resilient cities, as reported in the literature, focusing on the Global South. The analysis highlights the disproportionate application of resilience strategies across cities in the Global South. Next, we examine the Saudi Arabian context for analyzing urban resilience, focusing on the readiness for and strategies used by Saudi Arabian cities toward resilience building. These analyses show that, while some progress has been made toward urban resilience, there are significant barriers, particularly relating to climate change management, that obliterate the potential for resilient urban futures. Finally, we discuss the implications of the Saudi case within the broader literature.

5.2 Why Should Cities in the Global South Desire Urban Resilience?

Discourses on resilience remain ongoing, and have received considerable attention among scholars, governments, non-governmental organizations and international organizations (e.g., Abu-Rayash et al., 2023; Hosseini et al., 2016; Leichenko, 2011; Marana et al., 2019; Sharifi & Yamagata, 2016). And papers published in the *Sustainable Cities and Society Journal* have been central to urban resilience debates in recent times. For example, through these debates, an understanding of resilience has emerged in the context of city management guidelines (Marana et al., 2019), sustainability models (Abu-Rayash & Dincer, 2023), earthquake response (Kojima et al., 2014), community participation (Mahajan et al., 2022), and climate management (Croce et al., 2021). While these studies provide a context and foreground resilience thinking in urban development agendas and concerns, they do not necessarily demonstrate the urgency of resilience in cities in the Global South. This section therefore adds to the ongoing debates by demonstrating the critical importance of resilience paradigm in cities in the Global South. We briefly explain the meaning of resilience, and analyze its importance in climate change and urban planning focusing on cities in the Global South. This analysis is important in rethinking the framing of urban resilience in the Global South.

The term "resilience" is used to describe a system's capacity to tolerate a tragic event and either continue functioning normally or recover quickly from its effects (Leichenko, 2011). Cutter et al. (2013) define it as the capacity to anticipate, tolerate, recuperate from, and effectively adjust to stressful situations. In other words, resilient systems can quickly recover from disruptions and continue to function normally. This means that such systems require the ability to plan and prepare for a disaster, absorb its effects, and recover from the effects of a disaster. While engineering takes a static view of resilience, other professions, such as those concerned with social and ecological systems, take a more dynamic perspective, focusing on the ability to "adapt." This allows them to account for disruptions that may lead to various equilibrium or non-equilibrium conditions. Although the concept of urban resilience is novel, its definition remains ambiguous (Amirzadeh et al., 2022) and lacks consensus on the features that characterize it and the most suitable analytical unit for measuring it. For example, resilience is distinguished by several characteristics in fields such as engineering, including robustness, reliability, variety, multiplicity, resourcefulness, inventiveness, flexibility, adaptability, productivity, self-organization, inclusion, and the ability to foresee potential problems (Hosseini et al., 2016; Sharifi & Yamagata, 2016).

Several studies have been conducted on urban resilience evaluation (Salehi & Sharifi, 2022; Sharifi et al., 2021, 2022; Sharifi, 2022b). It is important to evaluate a wide range of characteristics specific to urban contexts before attempting to develop resilience in urban areas. However, only a small number of authors have considered how well this idea translates to other geographical contexts. Leichenko (2011) suggests that the many different fields of study that contribute to our understanding of urban resilience could be grouped into four main categories: urban ecological resilience, urban and regional economic resilience, urban hazards and disaster risk reduction, and the promotion of resilience through urban governance and institutions. These bodies of literature share significant common ground, but each highlights distinct aspects of urban resilience and concentrates on various aspects of urban systems and infrastructure. The United Nations' contribution toward achieving the 2030 Agenda for Sustainable Development in a risk-informed and collaborative way is the UN Plan of Action on Disaster Risk Reduction for Resilience. It is one of the most important resources that the United Nations offers to help countries improve their development planning, funding, and programming, considering the risks involved (UNDRR, 2023). The program outlines the principles, promises, directions, and goals for disaster loss reduction and for aiding governments and communities in implementing the Sendai Framework. This program involves aspects that address many different facets of society, including infrastructure, safety, environment, economics, and governance. For urban systems to be resilient, city planners must prioritize the organizational component, which requires hierarchical relationships between city leaders and the urban population. The urban environment, effectiveness and robustness of strategic services, and state of physical infrastructure all contribute to what is known as the spatial dimension (Alizadeh & Sharifi, 2020).

Multiple overarching themes of urban resilience have emerged in the context of climate change (Leichenko, 2011; Mahdaoui & Alibou, 2021). Mahdaoui and Alibou (2021) argue that expanding the use of renewable energy sources such as solar, wind, and nuclear energy, could help create cities that are more resilient to climate change and natural disasters. Reuse and recycling, separation of solid waste, downsizing of bags, packaging to limit garbage disposal into landfills, and many more strategies have received much attention in recent years (Amirzadeh et al., 2022; Dastjerdi et al., 2021). Metropolitan areas must be more adaptable to withstand the effects of climate change (El Mallah, 2019). For example, the Saudi Green Initiative is one of the country's national development frameworks and policy initiatives. The goals for 2030 set forth by the Saudi Green Initiative include increasing the share of renewable energy in energy production to 50%, reducing carbon emissions by more than 4%, planting 10 billion trees countrywide, and expanding protected areas to cover more than 30% of Saudi Arabia's total land area (Saudi Green Initiative, 2021). The United

Arab Emirates was the first Gulf country to implement a new energy strategy, which includes nuclear power, solar energy, and natural gas, to reach net-zero emissions by 2050 (Barrington, 2021). To improve the public's health and well-being and cut down on energy use, the government of Dubai published "Green Building Regulations" for the private sector. Buildings are allowed and are expected to reduce their energy and water usage by 30% as a result of this legislation (UAE Government Portal, 2022). In 2019, the construction of all new buildings followed this code. Some have made rudimentary plans for future emission reductions; for example, in 2022, Kuwait's cabinet ordered an examination of ways to reduce the country's production of greenhouse gases. This followed an earlier vow to trim emissions of merely 7.4% by 2035 (MacDonald, 2022). Outside the Gulf region, other countries have lauded climate-friendly initiatives. For instance, Morocco has emerged as a global leader in solar energy (Alami, 2021).

The mainstream literature on climate-resilient cities (Fig. 5.1) presents key theoretical underpinnings and empirical findings. Extra analytical examples of what makes a city resilient, elaborating on the significance of resilience in land-use and urban planning systems, disaster management, green infrastructural systems, digital technologies for climate resilience, climate-smart cities, resilient coastal economies, institutional structures, and public participation. These assumptions have been used to analyze how well Saudi Arabian urban areas can withstand the effects of climate change. Many people worry that climate change makes it more difficult for them to have the resources they need to live comfortably. These include clean water and air, medical care, good education, and a government voice. Therefore, land resource management, water and energy accessibility, and affordability should all be factored into development strategies alongside the promotion of economic expansion (Almulhim & Abubakar, 2023; Sharifi, 2020). Furthermore, these strategies should consider the potential future impact of climate change. To this end, this chapter builds on the generally accepted assumption that climate-resilient-development takes place once society seeks growth in the economy, alleviation of poverty, and other long-term goals, while also sequentially integrating the consideration of present and future climate threats into development strategies.

However, the cost of adopting a resilience strategy and inadequate scientific knowledge of the usefulness of urban policy are only two potential stumbling blocks that, according to Hamouda and Okumus (2019), resilience initiatives may face. It is essential to first assess the relevance of these techniques in building urban resilience, given that they are likely to have ramifications on how various cities address climate change. Risk, resilience, and adaptive ability rather than catastrophic damage should be prioritized in analytical efforts. Making systems more resilient and less vulnerable to climate change by taking measures to lessen the severity of adverse effects or increase their ability to adapt to these changes is what adaptation means (Mahdaoui & Alibou, 2021).

Urban planning and land-use systems	It is well acknowledged that planning systems are crucial for making cities more resilient. Proper urban and land-use planning and implementation systems foster urban resilience by fostering sustainable development, generating economic growth, and promoting environmental protection.	Asibey et al., 2022; Burby et al., 2000; Cobbinah & Finn, 2022; Leichenko, 2011; Mahdaoui & Alibou, 2021; Sharifi, 2022a
Disaster management	Damage to both people and property can result from urban flooding. Urban resilience can be improved by establishing systems for dealing with disasters, such as early warning and evacuation, first aid, ongoing aid, rapid infrastructure restoration, and more.	AlGahtani et al., 2022; Alrehaili, 2021; Ledra & Al-Ghamdi, 2020; Liao, 2012
Green infrastructural systems	Integrating green areas into the city's spatial structure improves urban dwellers' habitat, wellness, and health and is essential to urban sustainability.	Alizadeh & Sharifi, 2020; Infield et al, 2018; Meerow & Newell, 2017; Sturiale & Scuderi, 2019
Digital technologies embracing climate resilience	New digital tools have the potential to supply smart infrastructure that allows for smarter decision-making pre-, during-, and post-hazard incidence by providing real-time performance measures. Technologies such as IoT, BIM, digital twins, AR/VR, AI, and a host of others, are already being used in various industries. Nonetheless, their full potential has yet to be realized. By utilizing these resources and the data produced, we could facilitate the delivery and dissemination of automated, quick, and precise analyses to enhance the climate resilience of our public infrastructure.	Argyroudis et al., 2022; Balogun et al., 2020; Papa et al., 2015
Climate-smart cities and resilient coastal economies	Smart urban planning that considers climate change is essential for enhancing the quality of life in and stimulating the economy of coastal communities. Standards for disaster mitigation, green economy, and low-carbon transport are all part of climate-smart urban policies that should be implemented in vulnerable areas. Despite being economic hubs, urban and coastal regions are ground zero for climate change's effects. Carbon neutrality, resilient and inclusive cities, and coastal regions are possible with climate-smart spatial planning and crucial investments.	Kuo et al., 2022; Salehi & Sharifi, 2022; Sharifi, 2022b; Sharifi & Allam, 2022; Sharifi et al., 2021, 2022
Governance and public participation	Predicting and preparing for disruptions are clear indicators of urban resilience; these, in turn, depend on the technical capabilities of responsible institutions, the general performance of a country's systems, and its governance procedures. Furthermore, public participation is critical for the effective execution of climate actions. Social relationships, integration, and cooperation are the backbone of strong communities, and any plan to boost community resilience must capitalize on pre-existing links among members.	Burton & Mustelin, 2013; Carmin et al., 2009; Sharifi, 2020; Wamsler et al., 2013

Fig. 5.1 Concepts and theoretical premises for a climate-resilient city (Illustration created by the authors)

5.3 Analytical Context and Approach

5.3.1 Analytical Context

This analysis focuses on cities in Saudi Arabia, which is a Western Asian country. With a total geographic area of 2.15 million square kilometers (or approximately 80% of the Arabian Peninsula), Saudi Arabia ranks as the 14th largest country in the world, the largest in the Middle East, and the second largest in the Arab world (CIA.gov, 2023) (Fig. 5.2). The population density ranges from 118.7 people per square kilometer in the Jizan area to 2.9 people per square kilometer in the northern border region, with a national average of 12.8 people per square kilometer in 2010 (Gazzeh & Abubakar, 2018). The country's GDP of USD 1775 billion in 2017 places it within the world's top 20 best-performing economies, making it the wealthiest nation in the Middle East and North Africa (MENA) (CIA.gov, 2023).

Saudi Arabia was selected for this analysis for the following reasons: First, it has undergone a significant phase of rapid urban growth. Notably, Saudi Arabia saw the number of cities growing from 58 to 285 between 1936 and 2015, distributed over 13 regions and 118 governorates (UN-Habitat, 2016). According to data from 1950 to 2025, urbanization in the nation has increased exponentially from 21 to 83%, with 85% of the population residing in cities (Alarabi & Alasmari, 2023; Almulhim & Cobbinah, 2023a; Almulhim et al., 2022). Such rapid urbanization warrants an analysis of its implications for climate change. Similarly, one of the biggest urban planning problems in Saudi Arabia is poor coordination in urban planning due to the sharing of urban tasks among multiple organizations that are answerable to different authorities (Bakhit, 2023). Without proper coordination, especially in rapidly growing urban areas, addressing and implementing initiatives to tackle important issues such as climate change would be problematic. The Middle East is the world's richest oil-reserve region, with Saudi Arabia as its hub (Mahmood et al., 2020). However, the nation's reliance on oil makes it the world's largest greenhouse gas emitter (Fang & Lerner, 2019). The nation is likely to experience serious climatic changes if the high emissions continue. For example, without immediate action, the occurrence of agricultural droughts in the Saudi Arabia will increase by 88% by 2050 (G20 Climate Risk Atlas, 2021), and the nation will experience more severe and prolonged heatwaves. Adverse impacts affect both the physical environment and economy. Saudi Arabia is likely to witness a 12.2% decline in GDP over the next 30 years if no action on climate change is taken (Radwan, 2022). Therefore, it is essential that Saudi Arabia acts immediately to reduce its carbon footprint and become more resilient to climate change. Thus, Saudi Arabia offers a useful context for examining how cities respond to the critical challenges of climate change and rapid urban growth through resilience strategies.

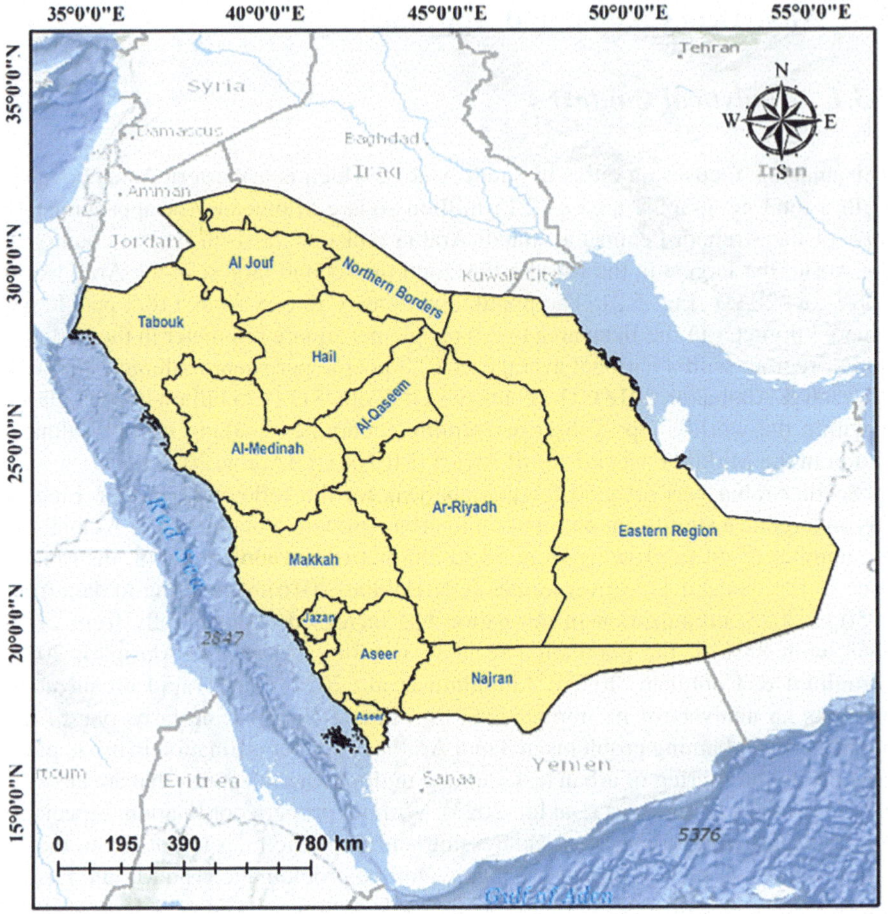

Fig. 5.2 Administrative divisions of Saudi Arabia within the regional context of the Middle East. *Source* Author's own work

5.3.2 Data Gathering and Analysis Approach

Using Preferred Reporting Items for Systematic Reviews and Meta-Analyses (PRISMA) framework, we reviewed both conceptual and empirical studies on urban resilience in the global South, focusing on Saudi Arabia. PRISMA framework is useful for methodically conducting transparent and replicable procedures for consolidating existing research findings on a particular topic (Page et al., 2021). In this chapter, we used secondary data to explore the framings of urban resilience, and understand how urban policy outcomes influence climate action in Saudi Arabian cities. The literature search was conducted using the Google Scholar search engine, and major electronic databases including Scopus, Web of Science (WoS), and ProQuest. These electronic databases are hubs of comprehensive bibliometric data on urban resilience, climate change, and planning research. Their application in this chapter offered a refined results based on our keywords search. Different concepts and phrases were used as search keywords to ensure that all relevant and various types of information on resilience, planning and climate change were collected. Prior to the search on electronic databases, literature search criteria were designed and included publications, books, book chapters, articles, and policy documents, in the English Language with a contextual focus on Saudi Arabia, and the start and end year of 2000 and 2023 respectively.

From June to August 2023, a search query was conducted using the following keywords: "urban resilience," "climate change," "Global South," "developing countries," "urban planning," "land use," "disaster management," "risk reduction," "green infrastructure," "governance." A total of 16,047 records were generated from the literature search across the four databases (Google Scholar = 5105, ProQuest = 3587, Scopus = 4895, and WoS = 2460). After duplication checks, 12,849 records were removed. Additional checks were carried out with their titles and keywords to establish their relevance to the chapter, resulting in the elimination of 2158 records. The remaining list of records ($n = 404$) was then assessed against the inclusion and exclusion criteria by focusing individually on the abstracts of the records (Table 5.1). This led to the removal of 281 records, remaining 123 records for the final selection and inclusion (Fig. 5.3).

Of the total of 123 selected publications, about 75% were journal articles ($n = 98$) and 25% consist of policy documents, conference papers, books, and book chapters ($n = 24$). Figure 5.4 shows the number of annual publications, with over 80% of publications published since 2015, showing increasing recent interest in urban resilience and climate change management issues. The analysis shows a gradual recognition of urban planning's role in climate change management, and resilient building in Saudi Arabia (approximately 60% of reviewed studies directly or indirectly discuss urban

Table 5.1 Inclusion and exclusion publication criteria

Inclusion criteria	Exclusion criteria
Climate change	Irrelevant to research aim and objectives
Global South	Not applicable to Arabian context
Developing countries	Non-global South-focused green infrastructure research
Green infrastructure	Non-global South-focused urban planning
Governance	
Disaster management	
Risk reduction	
Urban planning in Saudi Arabia	
Land use	
Urban resilience	
Relevant to research purpose	

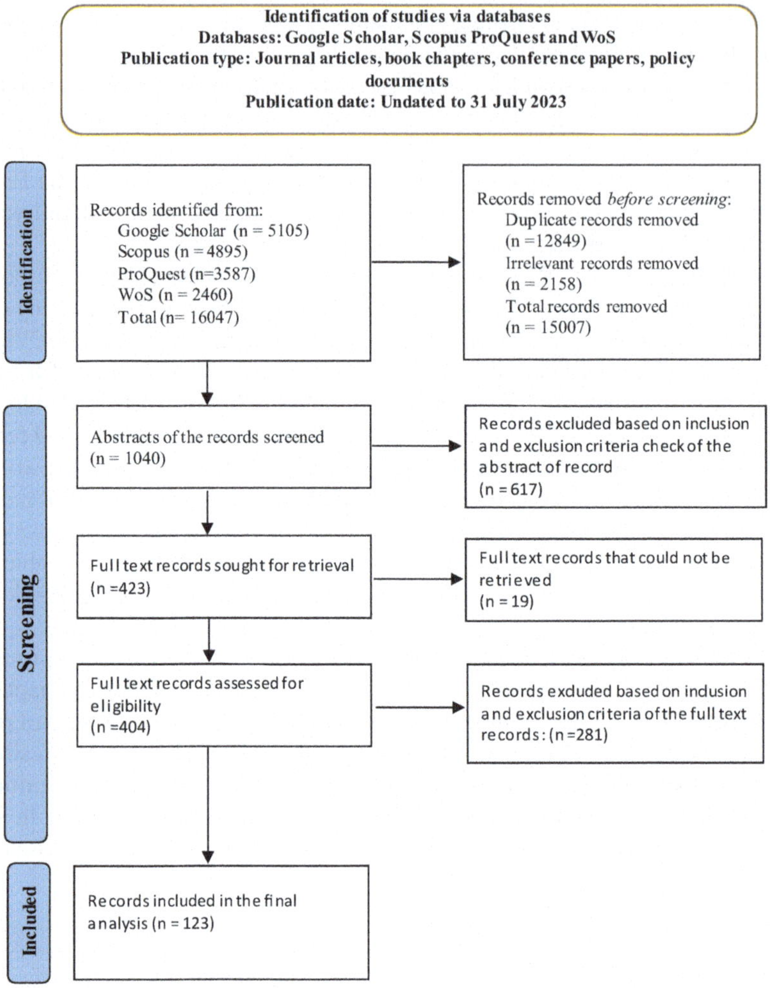

Fig. 5.3 PRISMA framework for analysis

planning and climate change management issues) in the context of improving urban planning response to urban mobility.

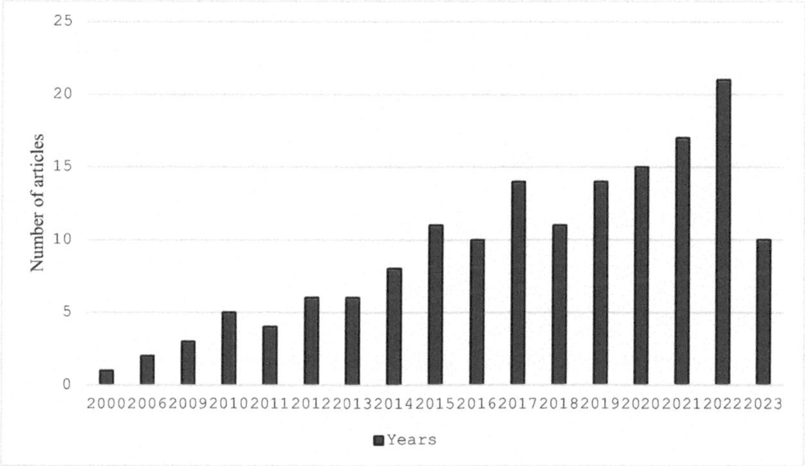

Fig. 5.4 Analysis of publication by year (2000–2023)

5.4 Urban Resilience in Saudi Arabia: Analysis of Climate Change and Urban Policy Readiness

5.4.1 Saudi Arabian Cities' Readiness for Future Uncertainty

In the last 30 years, 14 natural disasters have hit Saudi Arabia, harming approximately 30,000 individuals and costing about USD 450 million. Flooding is the most prevalent type of disaster (UN-Habitat, 2019). Flash floods, slope failures, heat waves, landslides, and urban island heat are natural disasters to which the Saudi Arabia is vulnerable (Fig. 5.5). Slope failures, especially rockfalls, are widespread in mountains (Abualnaja, 2011; Lam et al., 2016). Erosion, dust storms, droughts, shifting dunes, and salinization are prevalent natural hazards in arid terrain, contributing to the deterioration of agricultural fields (Al-Bassam et al., 2014). Extreme heat in the Arabian Peninsula in 2010 led to a record-breaking demand for electricity, with summertime highs in Jeddah reaching 52 °C. Three-quarters of Saudi Arabia is predicted to experience an upsurge in dryness by the end of the century, and the Saudi Arabia is forecasted to warm faster than the rest of the world by 2040 (Fig. 5.5). Figure 5.6 illustrates the annual mean extreme temperature over the 2030–2080 period, which increases within the range of −0.6 °C to +5 °C.

Higher temperatures are felt in congested urban areas owing to climate change and the lack of green open spaces. Dust storms are becoming more common, affecting agriculture and tourism and are already experiencing the effects of climate change. Field crops and fruit trees are predicted to suffer yield losses of 5–25%, along with alterations in the regularity, pattern, and intensity of precipitation (UN Habitat, 2019). There have been more recurrent bouts of severe rainfall and flash

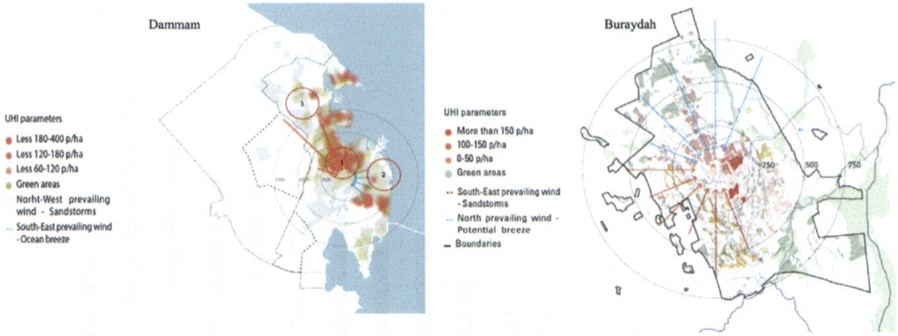

Fig. 5.5 Areas at risk of the urban heat island effect in Dammam and Buraydah cities. *Source* UN-Habitat (2019)

Fig. 5.6 Mean annual temperature trend (2030–2080) in Saudi Arabia. *Source* UN-Habitat (2019)

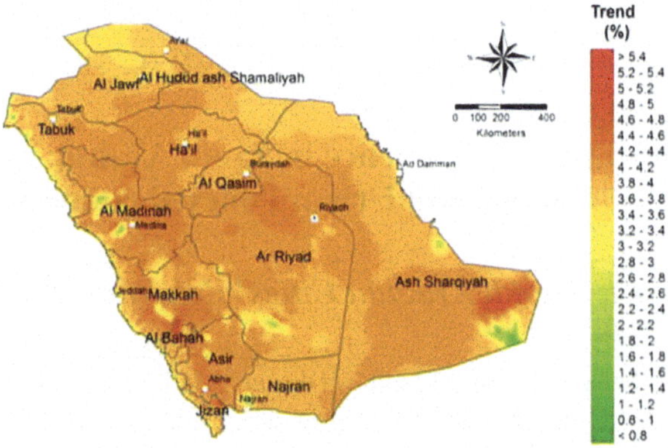

floods, although the monthly average rainfall fell from 6.24 mm to 6.06 mm between 2006 and 2009 and to 5.44 mm between 2010 and 2012 (Nahiduzzaman et al., 2015; UN-Habitat, 2019). Hence, rainfall would be restored to less than 50% of the water lost to evapotranspiration, leaving the surface water supply unchanged. As a result, municipal infrastructure, especially those that are weak or old, comes under increasing strain and eventually breaks down, putting the quality of already scarce water supplies at risk.

Over the past few decades, however, severe or unprecedented flooding or flash floods have emerged as the EM-DAT's frequently reported dangers of EM-DAT (CRED, 2021). Thousands of individuals in Saudi Arabia have been afflicted by flash floods in the last decade alone, resulting in enormous economic and human losses. Over 12% of Saudi Arabia's population live in low-lying coastal cities. Flash floods in Jeddah in 2009 and 2011 were the most catastrophic to Saudi Arabia over 30 years (Fig. 5.7). This was because of increased buildings in low-lying areas and wadis, as well as an increase in impermeable urban surfaces. Furthermore, strong storms that cause flash floods make unplanned city neighborhoods more susceptible to damage. Around 163 people were killed, hundreds were injured, thousands of people had to be rescued, and many thousands of homes and businesses were destroyed when a flash flood coursed through Jeddah City in 2009 (Alamri, 2010). This had a destructive effect on municipal water supplies, delicate ecosystems and wildlife ecosystems, manufactured structures, and urban greenery. Saudi Arabian cities are not built to endure climate change, and in response to this, the government has implemented laws and steps to make the country less susceptible to its impacts. Furthermore, 21% of Taif's unplanned areas are vulnerable to climate change, as are 26% of Makkas and 17% of Madinah's, and 32% of Taif's infrastructure requires further development, according to an evaluation of Saudi cities that showcases the necessity of developing and reinforcing urban resilience (UN-Habitat, 2019). Urban resilience necessitates cautious land-use planning, proactive environmental legislation that preserves ecological boundaries, safe construction methods, and a proactive attitude toward early warning systems.

While Saudi Arabia has made some important strides on the catastrophe risk management front, it should also prioritize creating a framework to back up local policies. Multiple publications, such as the Saudi Energy Efficiency Program (2012), Saudi Arabia's Vision 2030 (2016), the National Renewable Energy Program (2017), the Saudi National Water Strategy (2018), the Saudi Green Initiative (2021), and the Circular Carbon Economy National Program (2021), detail the policy provisions implemented in the country. For instance, in 2012, the Saudi Energy Efficiency Program was launched as an overarching framework for the Saudi Arabia's energy efficiency measures, targeting the buildings, industries, and transportation sectors that together account for 90% of Saudi Arabia's total energy demand. There are

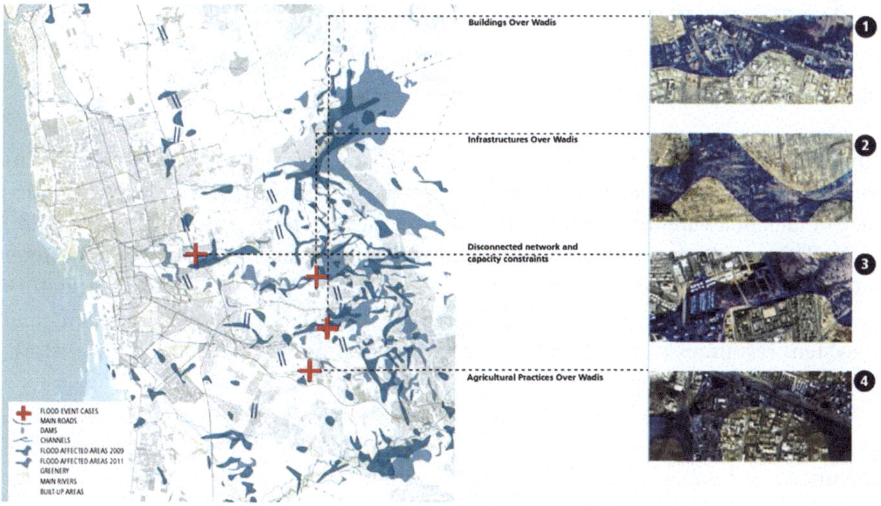

Fig. 5.7 Massive flood events in Jeddah 2009 and 2011. *Source* UN-Habitat (2019)

around 35 programs spread across 13 groups that aim to improve things such as public awareness, energy efficiency requirements, and benchmarks (Shehri et al., 2022). One of the many goals outlined in Saudi Arabia's Vision 2030 is the production of 9.5 GWh of renewable energy by 2023 (Khan, 2016). The vision calls for the eventual liberalization of the energy market and the localization of a large section of the renewable energy value chain within the Saudi economy, including R&D and production. There are plans to grow 10 billion trees and switch to renewable energy by 2030 as part of the Saudi Green Initiative (Hameed et al., 2022). The goal of Saudi Arabia's Circular Carbon Economy National Program is to make the features of the circular economy possible there. It prioritizes the application of cutting-edge technology in accordance with the 4Rs principle (Reduce, Reuse, Recycle, and Remove) (Shehri et al., 2022).

There is mounting evidence that floods are becoming more common in Saudi Arabian cities, and this trend is expected to continue unless appropriate action is taken. Flash floods killed over a thousand people in Saudi Arabia between 2000 and 2020, according to the International Disaster Database (EM-DAT) and caused billions of dollars' worth of economic losses of (Alrehaili, 2021). There is a wide range of disaster management emergency-preparation strategies across numerous countries. In the framework of the Saudi Arabia, the General Directorate of Civil Defence (GDCD) oversees emergency management and planning, as well as safeguarding human lives and material possessions (GDCD, 2020). However, the GDCD took a long time to develop its policies, laws, and regulations for emergency preparedness and disaster management despite its strong commitment to reducing risks and the severity of disasters (Abosuliman et al., 2013).

5.4.2 The Resilience of Saudi Arabian Cities

5.4.2.1 Land Use and Urban Planning

A major problem for people in Saudi Arabia and elsewhere is the negligent implementation of land-use planning restrictions, inadequate safeguards, and widespread ignorance or lack of awareness about the impact of waste and its management (Almulhim, 2022). This results from a lack of long-term planning and local engagement along with excessive government interference, which have led to the disposal of factory wastes and oil spills into water bodies, putting the lives of aquatic organisms and ecosystem resources delivered by surface water bodies in increased danger. Efforts are currently being undertaken to resolve these concerns.

Among these efforts is Saudi Arabia's National Center for Waste Management, whose stated goal is to control and oversee waste management operations, encourage investment, and enhance their quality in accordance with the concept of the circular economy in waste management to meet SDGs (MWAN, 2023). Institutions should recognize the significance of urban planning and land use because they could be used to protect communities from climate risks in many different ways. These include

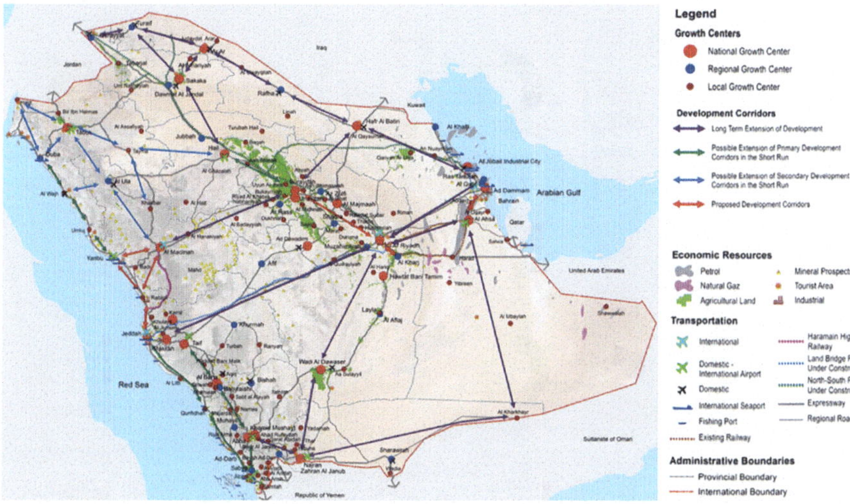

Fig. 5.8 National Spatial Strategy. *Source* Ministry of Municipal and Rural Affairs (2015)

restricting the development of urban areas, infrastructure, and economic activities in locations that are susceptible to natural hazards, ensuring that the urban environment is resilient enough to endure extreme levels of climate, and increasing the capacity of natural ecosystems to bounce back from disruptions. More than half of all human-caused greenhouse gas emissions occur in urban areas. Urban design, land-use planning, building regulations and height by-laws, transit planning, road networks, zoning, master plans, and subdivision regulations are all ways in which city and local governments could affect land- and energy-use processes. The National Spatial Strategy (see Fig. 5.8), developed under the direction of the Ministry of Municipal and Rural Affairs in 2001, emphasizes protecting the environment, limiting urban expansion, and promoting spatially balanced growth (UN-Habitat, 2019). On existing land use, Riyadh, Makkah, Dammam, and Al-Madinah are primarily distinguished by huge tracts of land with single-purpose uses, primarily residential, that extend from the city center across sizable distances. Most of these neighborhoods do not have high-quality open spaces, and access to services due to the availability of few mixed-use and commercial purposes land uses (Fig. 5.9). It is therefore essential to rebalance the development approach, to repair existing ecological damage, to protect and enhance natural assets in an urban context, and to consider current and future challenges (hybridization of climate change, coastal resilience, and the need for ecological corridors,) if necessary to create new mixed-use nodes.

5.4.2.2 Disaster Management

Studies reveal variations in preparedness owing to educational standards and the quantity of short courses on disaster risk management employees have finished. This

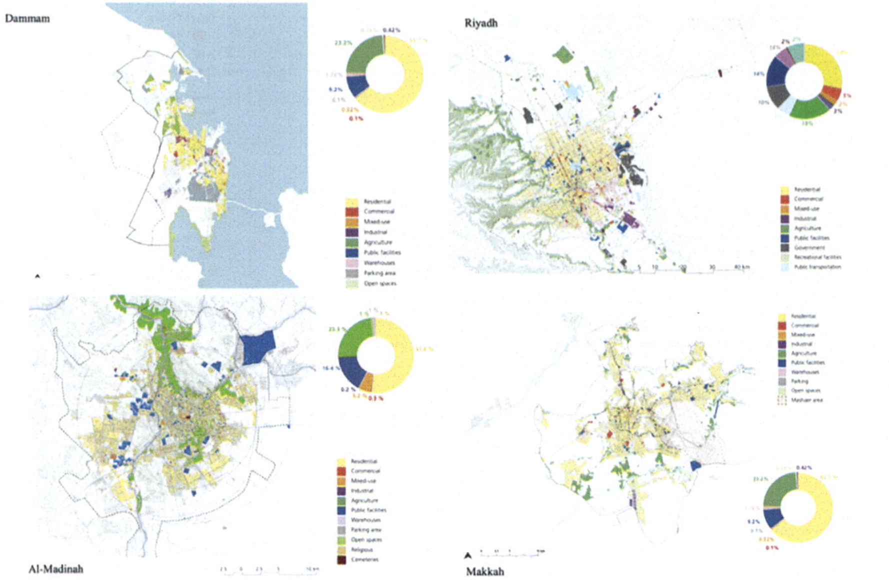

Fig. 5.9 Existing land use for some cites of Saudi Arabia *Source* UN-Habitat (2019)

emphasizes the necessity for organizations to think about enhancing their capacity and capability by means of pertinent educational programs (AlGahtani et al., 2022). Despite the GDCD's efforts, developing emergency planning and disaster management guidelines, procedures, and regulations has taken a long time due to a lack of staff and education, leaving its employees less prepared for potential future crises (Abosuliman et al., 2013; Alamri, 2010). The municipal civil defense department in Riyadh does not appear to pay a great deal of attention to its people's readiness for disaster management. The problem stems from a lack of adequate data collection and analysis following previous flash-flood tragedies in the city, as well as a lack of education on this subject (Ledraa & Al-Ghamdi, 2020). The procedure for managing flash floods ought to begin long before a regrettable incident happens. Reducing vulnerability to disaster consequences, such as injuries and property and life loss, is the goal of this phase. Regarding Riyadh, it appears that the municipal civil defense department places little emphasis on this stage of the process. Since these floods are unusual for the Riyadh, complacency often takes the place of worries about lessening their effects (Ledraa & Al-Ghamdi, 2020). This largely explains why there have not been any significant changes made to the local urban zoning laws or land-use regulations, such as the ones that forbid building in risky locations like low-lying lands (Ledraa & Al-Ghamdi, 2020). However, when they do occur, they are widespread with damaging and life threatening consequences. For instance, floods were reported in Jeddah City in 2009, 2010, and 2011. The projected monetary damages were $1 billion, with long installations and government facilities accounting for the majority of the costs. Compensation for those impacted was estimated to be $2 billion (Ameur, 2016). In 2009, there were 163 fatalities and 11,640 injuries (Abosuliman et al., 2013). When more floods hit Jeddah in 2011, it is estimated that 10,913 cars and 11,849 houses were damaged, and about 10 more people perished (Ameur, 2016). Riyadh recorded flood event in 2013 that led to some deaths and severe damages to properties and disruptions to city life and educational institutions (Ledraa & Al-Ghamdi, 2020). In addition, Makkah city recorded flood event in 2002 that led to 2 death and also Medina city flood recorded flood event in 2005 that led to 29 death and 43 Injured (Abosuliman et al., 2013).

Furthermore, heavy storms that drive flash floods increase the vulnerability of unplanned areas of cities such as Al-Madinah. As illustrated in Fig. 5.10, about 7% of the total cityscape of Al-Madinah remains unplanned, and about 70% of these unplanned regions are extremely vulnerable to flood events. This situation highlights the precarity of unplanned locations to climate change risks, especially flood events in Saudi Arabian cities.

The above explains to a significant extent why there are no major adjustments made to the local urban zoning and land-use management, for example, to prevent construction in risk-prone locations such as low-lying lands. There have been no changes made to the building codes that would strengthen structures so that they could better withstand the devastation caused by flash floods. These kinds of adjustments are essential in order to construct a city that is resistant to dangerous occurrences such as sudden flooding. Improving the resilience of Saudi Arabia's urban

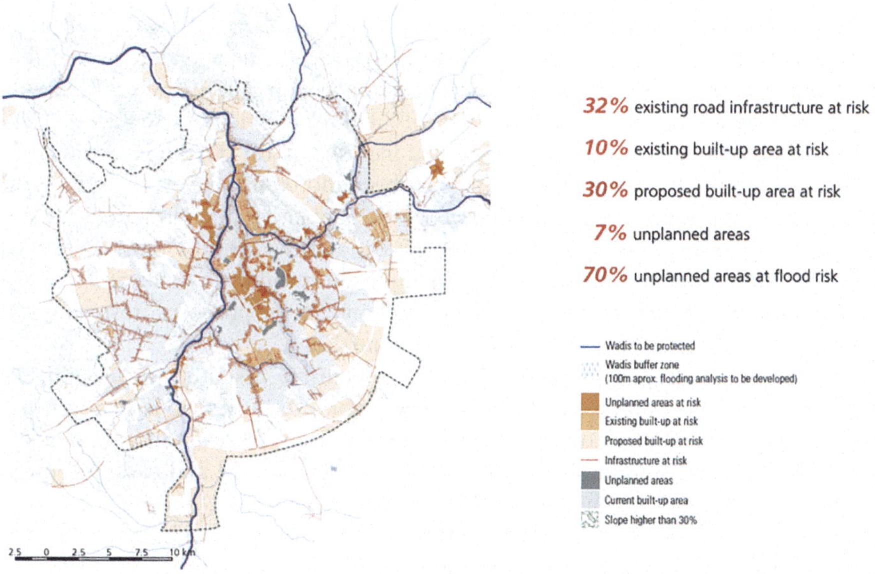

Fig. 5.10 Areas at risk of floods in Al-Madinah city. *Source* UN-Habitat (2019)

areas against disasters requires the development of institutions' capacities to provide stronger protection measures, dependable early warning systems, the introduction of flood insurance programs for businesses and residential properties, and efficient post-disaster cooperation for urban citizens.

5.4.2.3 Green Infrastructural Systems

Plans to enhance urban development in Saudi Arabia through the implementation of green, sustainable, and smart city initiatives and to increase citizen satisfaction are set to be unveiled as part of the transformation plans sourced from the National Transformation Program 2020 and Saudi 2030 Vision, as announced by the Saudi Ministry of Municipal and Rural Affairs (Alyami, 2019; USGBC, 2023); yet ecosystems and the green infrastructure they provide to cities are under constant threat due to growing urbanization and rising temperatures. Temperatures are expected to rise by 2–2.75 °C in coastal locations, and Al-Maamary et al. (2017) found that temperatures have increased by roughly 4 °C in Saudi cities over only five decades. In the summer, the air gets hotter than usual because of massive amounts of pollution from cars and other urban sources (Abubakar & Aina, 2018).

Urban areas have grown significantly, and city populations have increased rapidly over the past 40 years. A lack of infrastructural investment and eco-friendly approaches to city planning have hampered progress. In the case of Jeddah, for instance, the city's expansion has outstripped the capabilities of its infrastructure. The urban population of Saudi Arabia's other major cities, such as Riyadh and Dammam, has also exploded in recent decades (Addas & Maghrabi, 2021). Saudi Arabian cities, especially Riyadh, Jeddah, and Dammam, have a much lower green strategy (GS) availability per capita compared to international standards (Addas & Maghrabi, 2021). Despite climate change's prominence in Saudi Arabia, where the annual average temperature has risen by 0.75% and is projected to rise by 2.2 °C by 2050 compared to a global average projected increase of 1.5 °C, the important role of GSs in adapting to climate change has not been thoroughly investigated (Sharif, 2015). One of the Saudi Green Initiatives focuses on planting 10 billion trees nationwide (Fig. 5.11), the largest afforestation project and the world's most ambitious urban greening project. The purpose is to reduce temperatures by an average of 8–15 °C in afforested locations as well as for carbon capture purposes, restoring soil quality and enhancing the quality of life (Radwan, 2023).

Progressive GS implementation and management are essential for the citizens of Saudi Arabia's urban areas (Addas & Maghrabi, 2021). Previous research has proposed GS management options for GS preservation and restoration (Conedera et al., 2015; Govindarajulu, 2014; Jim, 2012). Carbon storage and carbon sequestration, according to Vaughn et al. (2014), can benefit from careful design. Because space for new GS is limited in urban areas, it is all the more critical to prioritize GS quality enhancement for the sake of recreation and biodiversity (Bolleter & Ramalho, 2014). Species introduction for a more natural GS composition is one example of how ecological principles could be utilized in design. Rooftop gardens, green walls, and

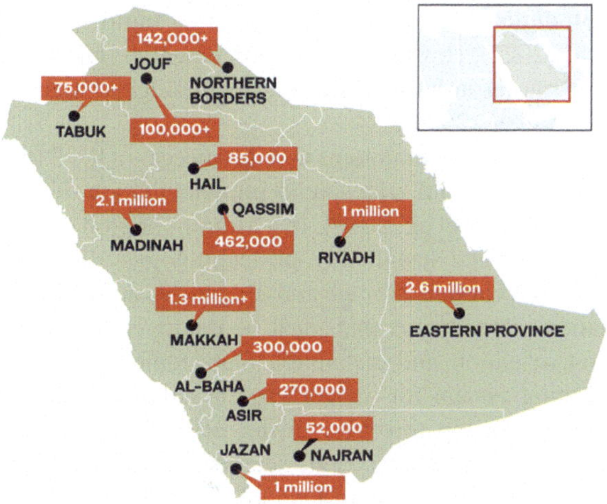

Fig. 5.11 Saudi Arabia's journey toward 10 bn trees (Radwan, 2023)

street sites are just a few examples of creative greening options that can improve GS quality (Jaafar et al., 2013). Redevelopment planning measures, such as conserving or relocating trees, can also be used to manage GSs to preserve the many direct and indirect benefits they provide. Vegetation can fill in the spaces where there once was none, such as in transit hubs with little room to spare.

Developing an eco-city model and implementing appropriate planning regulations is foundational to climate change management (Côté-Roy & Moser, 2022; Rapoport, 2014). Many different interpretations and implementations of the eco-city concept exist. That is to say, the eco-city might mean different things depending on the specific context in which it is located, as represented by urban initiatives and projects, and on the specific practices and techniques employed to attain its aims. Eco-cities, in the broadest sense, are human settlements that emphasize maintaining a natural environment and ecosystem that can support themselves (Rapoport, 2014). According to Bibri (2020), an eco-city district relies on renewable energy and data-driven technologies to meet the environmental goals of sustainable development in areas such as reduced energy consumption and pollution prevention. This includes using renewable energy sources such as wind, water, and sun; energy management; a bio-fueled combined heat and power (CHP) system; responsible waste management; a large-scale smart power source; and solar passive structures to reduce energy use and demand.

5.4.2.4 Digital Technologies Promoting Climate Resilience

There are real benefits to the resource and infrastructure system when assets are prepared for and adapted to specific threats using digital technology (Aerts et al., 2014). The development of a more resilient infrastructure to limit and recuperate from numerous hazards sustainably depends on the widespread implementation of new digital technologies supported by multi-stakeholder partnerships (Argyroudis et al., 2022). However, examining essential infrastructure's resilience to diverse and dynamic hazards through digital technologies has received surprisingly little attention (Koks et al., 2019).

Saudi Arabia maintains that the success of the Paris Agreement hinges on the rapid development and widespread implementation of cutting-edge technologies (Kane, 2021; Wogan et al., 2019). In addition, for global climate change goals to be reached efficiently and sustainably, all technologies should be considered. Carbon capture, direct-air capture, storage and utilization, and clean hydrogen are all promising mitigation technologies that must be deployed. Water conservation, recycling, collection, irrigation, and sustainable management for agricultural uses, as well as an early warning system for meteorological extreme occurrences such as floods and sandstorms, are top priorities in the field of climate adaptation technology. Both the governmental and corporate sectors in Saudi Arabia emphasize global collaboration and the dissemination of technology, and it is a driving force behind initiatives like Clean Energy Ministerial, Mission Innovation, and the Global Methane Initiative (Shabaneh et al., 2022). Saudi Arabia's latest smart city endeavor, ALNAMA, is planned to become a smart zero-carbon destination in the capital's outskirts of Saudi Arabia, Riyadh (Fig. 5.12). The city will promote sustainable living in a community of 10 square kilometers with various hubs including residential, educational, commercial, tourism, and medical (URB, 2022). The project is designed to adhere to the highest sustainability standards by offering multifunctional comprehensive solutions that address the three main pillars of sustainability: social, economic, and environmental. The goal is to build a smart city with a zero-carbon footprint that balances these three pillars. The initiative aims to establish a green circular economy for the city, which will be run entirely on renewable energy, recycled water, and recycled garbage (URB, 2022).

More stringent and data-driven approaches and tools for quantifying resilience need to be developed to realize the digital transformation of critical infrastructure. Argyroudis (2022) argues that emerging digital technology may help to address these problems. Such evaluations could be automated using data aggregation, digital and sensor technologies, and advanced simulation models powered by the Internet of Things (IoT), Artificial Intelligence (AI), Machine Learning (ML), and Building Information Modeling (BIM), among other things. With these tools, the architectural, engineering, and construction sectors may play their part in combating climate change by improving urban infrastructure. Improvements in productivity, quality of resources, stakeholder involvement, rework and waste management, sustainability and circularity, waste elimination in design, and organizational resilience are just a few of the many advantages of the concerted practice that BIM provides (Cube, 2021). Digital twins are another example of how AI is being used to automate and intelligently identify structural deterioration and destruction, as well as to predict the dynamic and adaptive effects of hazards on infrastructure (Argyroudis et al., 2022).

Nonetheless, their full potential has yet to be realized. Saudi Arabia must strengthen its lifecycle cost–benefit analysis for adaptation projects and its holistic

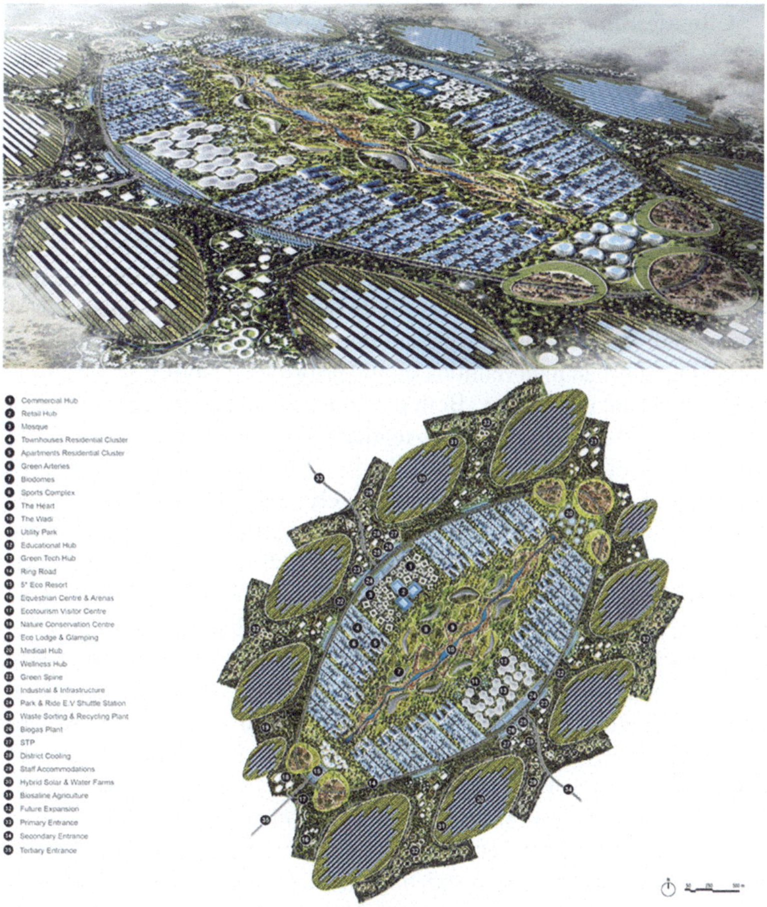

Fig. 5.12 ALNAMA Smart City masterplan (URB, 2022)

norms and legislation to make vital infrastructure more resilient through digital technologies (Aerts et al., 2014). Further study is needed in Saudi Arabia to construct adaptive capabilities and legislation supported by digital technologies that consider the profound vulnerabilities connected to climate change. Over 50 global models exist for predicting the course of climate change, but these predictions typically come with high uncertainty that can barely be modeled as random variables. There are unanswered problems regarding the effective application of digital technologies due to the need for simplified methods for combining data-based insights with the expertise of human inspectors to arrive at sound conclusions.

5.4.2.5 Climate-Smart Cities and Resilient Coastal Economies

Cities and coastal regions are on the frontline of climate impacts. Cities and coastal regions may become resilient and inclusive via climate-smart spatial planning and significant investments. The resilience of coastal economies can be improved by measures for Disaster Risk Management, such as the blue economy, climate adaptation, blue carbon sequestration, green growth, low-carbon transportation, and climate-smart urban plans (UN-Habitat, 2022a). Green buildings through innovation, green retrofitting, modifications for planning and building regulations, and the use of green technologies are a few examples of climate-smart urban strategies and innovations that Saudi Arabia must design and implement (UN-Habitat, 2022a, 2022b). Spatial planning of urban interventions should prioritize livability, green spaces, data utilization, and digital technology. Both government- and community-level support, planning, and execution are required to realize these objectives.

The population of the Arabian Gulf, both in cities and in the countryside, has steadily increased in recent years. Unfortunately, crucial natural ecosystems have been lost or severely degraded due to the unparalleled rate and magnitude of economic expansion. Jeddah is another city where septic tanks and unregulated industrial discharges are polluting groundwater systems (UN-Habitat, 2019). Untreated wastewater discharge threatens marine life and makes bathing areas unsafe. There has been a lot of dredging and filling done in the coastal areas of other important cities such as Yanbu, Jizan, and Jubail because of the growth of related sectors such as energy and desalination plants, waste disposal, and tourism. Coastal areas are under growing threat from the effects of climate change, including floods, erosion, and the introduction of salty water into ecosystems (UN-Habitat, 2019).

Saudi Arabia has lately developed privately motivated "eco-" or "smart cities," which often highlight private-sector design and engineering expertise for sustainable urban development, mirroring existing global climate change discussions. Eco-cities offer a chance to motivate urban development and restoration through socio-technological advancement, corporate development, and cultural brand recognition in the face of competing needs to reduce carbon emissions from cities as a central component of GHG emissions and to expand urban centers to satisfy socio-economic demands (Vishwa, 2021). Some new greenfield city developments underbuilt in Saudi Arabia to encourage industrial growth and diversify the country's oil-dependent economy are collectively called King Abdullah Economic City. Because of a newly available wealth of resources, the city's planners have prioritized water conservation, waste management, biodiversity preservation, and pollution control issues. Another smart city is NOEM City (Fig. 5.13) designed around automated driving aid passenger drones, repetitive tasks handled by robots, novel food growing and processing, free world-class online education, cities powered by renewable energy, 100% desalinated water using renewable energy to guarantee zero CO_2 emissions, AI and AR/VR technologies, full-scale e-governance, biotech innovation, wireless high-speed Internet, and healthcare technologies (Credable, 2022). The NEOM project is an important part of the Saudi Arabian Vision 2030, and a future megacity located in the Red Sea

Desert. It is approximately 33 times larger than New York City and is anticipated to cost $500 billion to construct (Credable, 2022).

5.4.2.6 Governance and Public Participation

Local emergency committee (LEC) members come from Saudi Arabia's many agencies (the police, the military, and internal security), but there is no space for community members. While the GDCD does keep international findings in mind, it does not provide a mechanism to ensure local community public participation, as is clearly outlined in international studies from a management standpoint (Alrehaili, 2021). The official mandate of the GDCD perceives participation solely in the context of persons volunteering during crisis-response emergency management. The public have only recently been activated as volunteers, allowing them to participate in rescue operations linked to disasters. Their function is reduced to that of simple respondents. Volunteers in the impacted areas of Saudi Arabia have been of significant help in recent disaster relief efforts, and their efforts have been well recognized. The government views itself as the sole protector of the state, and hence, it has yet to make many provisions for citizen involvement. Planned actions and policy decisions are made at the highest levels of management. In the context of disaster management, for example, this management style appears inappropriate due to the high demand for agility, quality output, rapid coordination, and efficacy in these settings (Ledraa & Al-Ghamdi, 2020).

The GDCD takes a rigid, top-down approach that relies heavily on a small number of people, whereas disaster response calls for a bottom-up approach to decision-making, a high degree of adaptability, and the transfer of responsibility to those on the ground. The time it takes for higher-ups to make decisions can be too long of a delay for rescue workers and other persons dealing with a disaster (Ledraa & Al-Ghamdi, 2020). For disaster relief to be effective, particularly for the most vulnerable communities, the participation of local groups and organizations is essential (Tambo et al., 2017; Van Krieken et al., 2017). Members of the local community know the area, the people who live there, the most pressing needs of the vulnerable population, and the best routes to get there. Tragedy could be adequately handled by an administrative structure with the active engagement of the community itself and increased public awareness of disaster preparedness (AlQahtany & Abubakar, 2020). The management of natural disasters is an obligation shared by all members of society, not just the government or a select group of organizations. Given their frequent occurrence, local communities have amassed a wealth of information and expertise in dealing with flash floods. This demonstrates how crucial it is to involve locals throughout each stage of disaster management, including risk assessment, advance warning systems, action prioritization, plan development, plan implementation, and plan evaluation to ensure mutually agreeable measures and strategies while fostering a shared feeling of accountability for reducing exposure to danger.

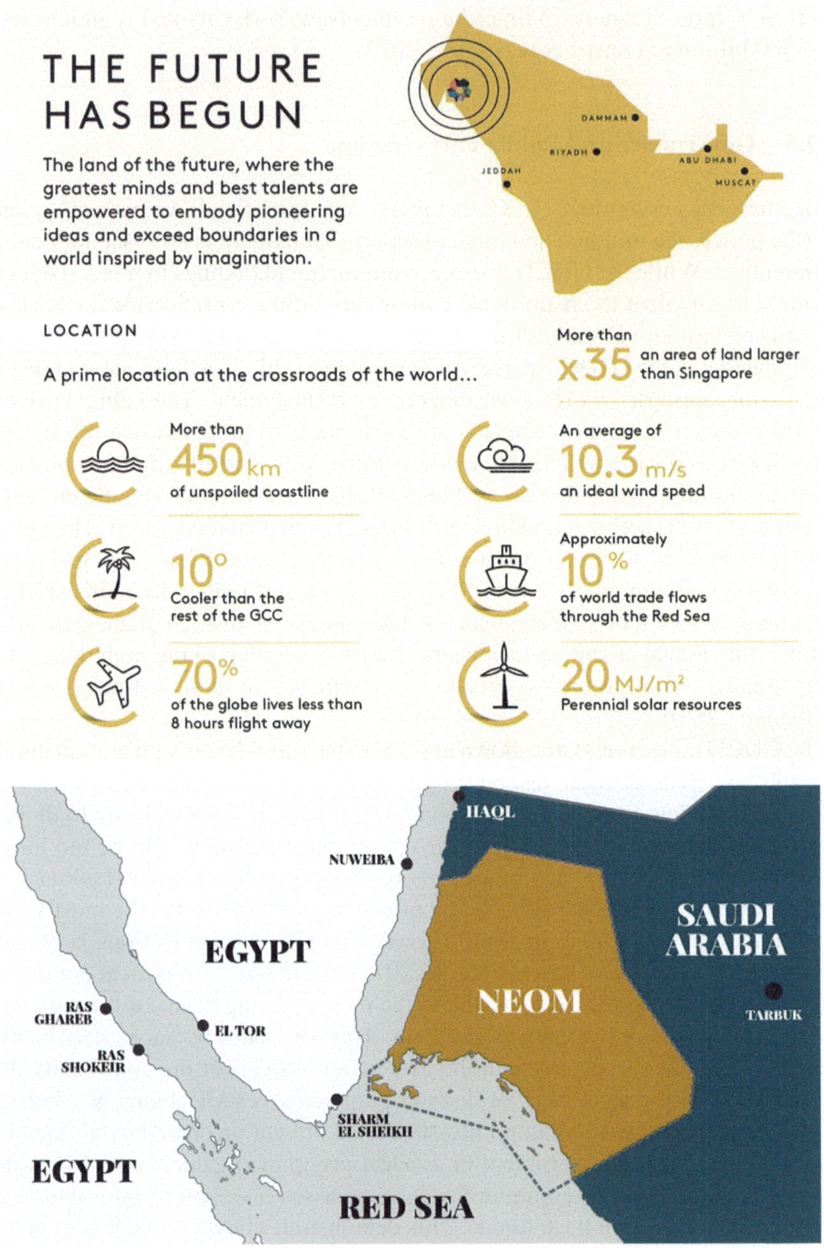

Fig. 5.13 NEOM is a planned cross-border city in the Tabuk Province of northwestern Saudi Arabia. It is planned to incorporate smart city technologies and function as a tourist destination (Credable, 2022)

5.5 Key Contributions

This chapter makes three primary contributions to understanding the relationship between climate change and urban policy within the context of resilience building. First, by focusing on urban policy and climate change, we have developed a novel approach of comprehensively assessing the colliding force of urban development and climate change on cities' resilience in the global south. Previous research across global south countries have predominantly focused on either climate change influences (e.g., Shehri et al., 2022) or urban policy/planning limitations (e.g., Abubakar & Aina, 2018; Almulhim & Cobbinah, 2023a) to urban functionality. Compared to them, this chapter extends beyond the current narrative and emphasizes the expression of urban resilience patterns dictated and framed by both climate change and urban policy, because these two global forces are simultaneously having significant impact on the urban landscape in the global south.

Second, our analysis provides a novel insight into climate change, urban policy, and resilience in the Arabian context (Abosuliman et al., 2013; Abualnaja, 2011; Ameur, 2016). Although previous studies have highlighted the severity of climate change risk in the Arabian context, it still remains to be demonstrated the importance of a nation's wealth to managing climate change risks. Our analysis shows that climate change risks are not necessarily less in wealthy global south countries. Despite Saudi Arabia's oil wealth, climate change risks are considerable (Fang & Lerner, 2019). Our analysis adds to the debate on climate change by demonstrating that without an urban policy agenda that understands, engages with, and reflects the aspirations of local communities, the risks of climate change may increase. In this analysis, an understanding of urban policy that fosters a shared feeling of accountability was established to be fundamental in managing climate change risks.

Finally, our analysis demonstrates the value of smart and innovative urban policy agenda in uncovering cities' responses to climate change and related urban development challenges. In previous studies, smart city innovations have been discussed as potential future of cities by analyzing issues of safety, climate change mitigation and highlighting the influence of technology in shaping the urban experience (Balogun et al., 2022; Jim, 2013; Zhou et al., 2021). However, these existing analytical frameworks fail to account for current climate change experiences and pressures in vulnerable cities in terms of adaptation (Hameed et al., 2022), while they are generally limited across the global south. Our analysis uses the smart and eco-city framework in Saudi Arabia to demonstrate the potential of managing climate change risks via urban policy, and further shows the limitations embodied in these emerging innovative frameworks. This is an important contribution to the climate change-urban policy dialogue in the global south (see Wang, 2022), and provides a context for reference and consideration by countries in their efforts toward building resilient futures.

5.6 Discussion and Conclusion: Desiring Resilience in Rapidly Urbanizing Saudi Cities

Projections of future climate change impacts and urbanization show an alarming risk of an uncertain future for many, if not most, cities across the Global South (UN-Habitat, 2019). The planning and management of urban spaces thus represents one of the most compelling and serious challenges cities in the Global South must address. Current debates and efforts aimed at addressing the threats of climate change and rapid urbanization in cities in the Global South focus mostly on building resilient cities through an atomistic agenda of addressing either climate change impacts or rapid urbanization, alongside progressive international development agendas (e.g., Sustainable Development Goals) (Cobbinah & Finn, 2022; Dastjerdi et al., 2021; Ochoa et al., 2018; Salehi & Sharifi, 2022; Sharifi, 2020). Yet such measures are insufficient to address ongoing and future urban development crises because they mostly fail to consider a consolidative and holistic policy agenda that responds to both urbanization and climate change pressures.

First, as we have shown here, urban resilience strategies relying solely on either climate change or rapid urbanization are counterproductive, as they limit the opportunities for a comprehensive planning and management agenda while increasing the footprint of climate change or urbanization impacts. Secondly, as shown by the analysis presented here, even when targeted at a comprehensive agenda to resilience building, inadequate analysis of the complexity and diversity of variables shaping and framing urban resilience at the city level would prove ineffective in terms of addressing the pressures of climate change and rapid urbanization. For example, in Saudi Arabia, green space in flood-prone locations, effective advance warning, and disaster-response mechanisms (including simple transportation access and public education concerning the effects of climate change and how to respond to it), are all efforts toward building resilience. However, the absence of efficient preventative, preparedness, and reduction initiatives within a disaster management system has resulted in thousands of unnecessary deaths. In view of this, urban resilience strategies should shift away from reactive approaches based on the idea that climate impacts are episodic. Instead, more proactive urban resilience policies are needed to recognize and address the short- and long-term urban development and management challenges that threaten the survival of cities in the Global South.

Our analysis shows that urban resilience in cities of the Global South can be triggered by the sustainable development and management patterns of land use, governance, and disaster management and the application of emerging technologies in urban planning (Alizadeh & Sharifi, 2020; Infield et al., 2018; Meerow & Newell, 2017; Sturiale & Scuderi, 2019). The Saudi case shows that long-term emergency planning and disaster management, which are more reactive to present natural hazards and less prepared for possible future catastrophes, have undermined the GDCD's ability to build resilience and protect Saudi Arabians from disasters. Critical urban studies and the human geography literature explain that these patterns are produced by distinctive and conscious economic and political structures that seek proactive sustainable development agendas and target the growth and well-being of vulnerable locations within the urban spaces (Alrehaili, 2021; Ledraa & Al-Ghamdi, 2020; Mahdaoui & Alibou, 2021). In other words, there is nothing natural about climate change or rapid urbanization impacts threatening and endangering the survival of cities, especially the most vulnerable areas. Instead, the severity of climate change threats and rapid urbanization challenges in cities in the Global South are products of inadequacies of governance, planning, and policy (Carmin et al., 2009; Ledraa & Al-Ghamdi, 2020; Wamsler et al., 2013).

To conclude, debates on urban resilience suggest that the only way to reverse the unsustainable and uneven patterns of urban vulnerability in cities in the Global South is by re-imagining an urban space where the challenges of rapid urbanization and climate change are fully analyzed and embedded in the urban policy agenda, and a proactive urban governance regime prevails. Our analysis confirms that the only way to build resilience in cities in the Global South, particularly in Saudi Arabia, is to strengthen land-use planning, make governance more people-centered, optimize the use water for amenities, and embed innovation and technology in urban policy actions. The challenge with such actions is that they are in stark opposition to the prevailing political hegemony across many cities of the Global South, built on over-concentration and over-exploitation of natural resources marked by inconsequential efforts to address climate change and the challenges of rapid urbanization. While decongestion measures, affordable housing, and regulation of the speed, volume, reliability, and placement of construction are all being pursued in Saudi Arabia, our analysis shows that city dwellers' vulnerability has been compounded by unchecked urbanization and rapid population expansion. We argue for a reorientation of current urban policy in Saudi Arabia toward a new environmental paradigm that progresses a proactive and inclusive agenda to address climate change and rapid urbanization. As UN-Habitat (2019) points out, the fight against climate change must include a system of governance that operates on multiple levels.

Acknowledgements This chapter is based on the following previously published article and is reproduced with permission from Elsevier: Almulhim, A. I., & Cobbinah, P. B. (2024). Framing resilience in Saudi Arabian cities: On climate change and urban policy. Sustainable Cities and Society, 101, 105172.

References

Abosuliman, S. S., Kumar, A., Alam, F., & Rasjidin, R. (2013, January). Disaster planning and management in Jeddah, Saudi Arabia. In *Proceeding of 2013 International Conference on Economics and Social Science* (pp. 20–21).

Abu-Rayash, A., & Dincer, I. (2023). Development of an integrated sustainability model for resilient cities featuring energy, environmental, social, governance and pandemic domains. *Sustainable Cities and Society, 92*, Article 104439.

Abualnaja, Y. O. (2011). Sea-level activities in the Kingdom of Saudi Arabia. *GLOSS GE, 12.*

Abubakar, I. R., & Aina, Y. A. (2018). Achieving sustainable cities in Saudi Arabia: Juggling the competing urbanization challenges. In *E-planning and collaboration: Concepts, methodologies, tools, and applications* (pp. 234–255). IGI Global.

Addas, A., & Maghrabi, A. (2021). Role of urban greening strategies for environmental sustainability—A review and assessment in the context of Saudi Arabian megacities. *Sustainability, 13*(11), 6457.

Aerts, J. C., Botzen, W. W., Emanuel, K., Lin, N., De Moel, H., & Michel-Kerjan, E. O. (2014). Evaluating flood resilience strategies for coastal megacities. *Science, 344*(6183), 473–475.

Ameur, F. (2016). Floods in Jeddah, Saudi Arabia: Unusual phenomenon and huge losses. What prognoses. *E3S Web of Conferences, 7*, 04019. https://doi.org/10.1051/e3sconf/20160704019

Alami, A. (2021). *How Morocco went big on solar energy*. BBC News. Available online https://www.bbc.com/future/article/20211115-how-morocco-led-the-world-on-clean-solar-energy

Alamri, Y. A. (2010). *Emergency management in Saudi Arabia: Past, present and future* (p. 21). University of Christchurch.

Alarabi, S., & Alasmari, F. (2023). Challenges and opportunities for small cities in the Kingdom of Saudi Arabia: A study of expert perceptions. *Sustainability, 15*(8). https://doi.org/10.3390/su15086960

Al-Bassam, A. M., Zaidi, F. K., & Hussein, M. T. (2014). Natural hazards in Saudi Arabia. *Extreme Natural Events, Disaster Risks and Societal Implications*, 243–251.

AlGahtani, A., Al Momani, N., Davies, A. J., & Alam, E. (2022). Identifying demographic, social and professional characteristics for effective disaster risk management—A case study of the Kingdom of Saudi Arabia. *Sustainability, 14*(22), 15399.

Alizadeh, H., & Sharifi, A. (2020). Assessing resilience of urban critical infrastructure networks: A case study of Ahvaz, Iran. *Sustainability, 12*(9), 3691.

Al-Maamary, H. M., Kazem, H. A., & Chaichan, M. T. (2017). The impact of oil price fluctuations on common renewable energies in GCC countries. *Renewable and Sustainable Energy Reviews, 75*, 989–1007.

Almulhim, A. I. (2022). Households' awareness and participation in sustainable electronic waste management practices in Saudi Arabia. *Ain Shams Engineering Journal, 13*(4), Article 101729.

Almulhim, A. I., & Cobbinah, P. B. (2023a). Can rapid urbanization be sustainable? The case of Saudi Arabian cities. *Habitat International, 139*(102), 884. https://doi.org/10.1016/j.habitatint.2023.102884

Almulhim, A. I., & Cobbinah, P. B. (2023b). Urbanization-environment conundrum: An invitation to sustainable development in Saudi Arabian cities. *International Journal of Sustainable Development & World Ecology, 30*(4), 359–373. https://doi.org/10.1080/13504509.2022.2152199

Almulhim, A. I., & Abubakar, I. R. (2023). Developing a sustainable water conservation strategy for Saudi Arabian cities. *Groundwater for Sustainable Development, 23*, Article 101040. https://doi.org/10.1016/j.gsd.2023.101040

Almulhim, A. I., Bibri, S. E., Sharifi, A., Ahmad, S., & Almatar, K. M. (2022). Emerging trends and knowledge structures of urbanization and environmental sustainability: A regional perspective. *Sustainability, 14*(20), 13195. https://doi.org/10.3390/su142013195

Alrehaili, N. (2021). An investigation into emergency planning requirements and challenges of disaster management in the Kingdom of Saudi Arabia. *International Journal of Disaster Management, 4*(3), 1–10.

AlQahtany, A. M., & Abubakar, I. R. (2020). Public perception and attitudes to disaster risks in a coastal metropolis of Saudi Arabia. *International Journal of Disaster Risk Reduction, 44*, Article 101422.

Alyami, S. H. (2019). Opportunities and challenges of embracing green city principles in Saudi Arabia future cities. *IEEE Access, 7*, 178584–178595.

Ameen, R. F. M., & Mourshed, M. (2017). Urban environmental challenges in developing countries—A stakeholder perspective. *Habitat International, 64*, 1–10.

Amirzadeh, M., Sobhaninia, S., & Sharifi, A. (2022). Urban resilience: A vague or an evolutionary concept? *Sustainable Cities and Society*, 103853.

Argyroudis, S. A., Mitoulis, S. A., Chatzi, E., Baker, J. W., Brilakis, I., Gkoumas, K., & Linkov, I. (2022). Digital technologies can enhance climate resilience of critical infrastructure. *Climate Risk Management, 35*, Article 100387.

Asadzadeh, A., Kötter, T., Fekete, A., Moghadas, M., Alizadeh, M., Zebardast, E., & Hutter, G. (2022). Urbanization, migration, and the challenges of resilience thinking in urban planning: Insights from two contrasting planning systems in Germany and Iran. *Cities, 125*, Article 103642.

Asibey, M. O., Mintah, F., Adutwum, I. O., Wireko-Gyebi, R. S., Tagnan, J. N., Yevugah, L. L., & Abdul-Salam, A. J. (2022). Beyond rhetoric: Urban planning-climate change resilience conundrum in Accra, Ghana. *Cities, 131*, Article 103950.

Bakhit, W. I. (2023). *Urban planning trends and challenges in the pursuit of Saudi Vision 2030*. King Salman Center for Local Governance. Available online http://www.ksclg.org/en/publication-project/urban-planning-trends-and-challenges-in-the-pursuit-of-saudi-vision-2030/

Balogun, A. L., Marks, D., Sharma, R., Shekhar, H., Balmes, C., Maheng, D., & Salehi, P. (2020). Assessing the potentials of digitalization as a tool for climate change adaptation and sustainable development in urban centres. *Sustainable Cities and Society, 53*, Article 101888.

Balogun, A. L., Adebisi, N., Abubakar, I. R., Dano, U. L., & Tella, A. (2022). Digitalization for transformative urbanization, climate change adaptation, and sustainable farming in Africa: Trend, opportunities, and challenges. *Journal of Integrative Environmental Sciences, 19*(1), 17–37.

Barrington, L. (2021). *UAE launches plan to achieve net zero emissions by 2050*. Reuters. Available online https://www.reuters.com/world/middle-east/uae-launches-plan-achieve-net-zero-emissions-by-2050-2021-10-07/

Bautista-Puig, N., Benayas, J., Mañana-Rodríguez, J., Suárez, M., & Sanz-Casado, E. (2022). The role of urban resilience in research and its contribution to sustainability. *Cities, 126*, Article 103715.

Bibri, S. E. (2020). The eco-city and its core environmental dimension of sustainability: Green energy technologies and their integration with data-driven smart solutions. *Energy Informatics, 3*(1), 1–26.

Busayo, E. T., Kalumba, A. M., & Orimoloye, I. R. (2019). Spatial planning and climate change adaptation assessment: Perspectives from Mdantsane Township dwellers in South Africa. *Habitat International, 90*, Article 101978.

Bolleter, J., & Ramalho, C. E. (2014). The potential of ecologically enhanced urban parks to encourage and catalyze densification in greyfield suburbs. *Journal of Landscape Architecture, 9*(3), 54–65.

Boyd, E., & Juhola, S. (2015). Adaptive climate change governance for urban resilience. *Urban Studies, 52*(7), 1234–1264.

Burby, R. J., Deyle, R. E., Godschalk, D. R., & Olshansky, R. B. (2000). Creating hazard resilient communities through land-use planning. *Natural Hazards Review, 1*(2), 99–106.

Burton, P., & Mustelin, J. (2013). Planning for climate change: Is greater public participation the key to success? *Urban Policy and Research, 31*(4), 399–415.

Carmin, J., Roberts, D., & Anguelovski, I. (2009, June). Planning climate resilient cities: Early lessons from early adapters. In *Fifth Urban Research Symposium, Cities and Climate Change: Responding to an Urgent Agenda* (pp. 28–30).

Centre for Research on the Epidemiology of Disasters (CRED). (2021). *EM-DAT database*. Available online https://www.emdat.be/database

Champlin, C., Sirenko, M., & Comes, T. (2023). Measuring social resilience in cities: An exploratory spatio-temporal analysis of activity routines in urban spaces during COVID-19. *Cities, 135*, Article 104220.

CIA.gov. (2023). *Saudi Arabia*. The World Factbook. Available online https://www.cia.gov/the-world-factbook/countries/saudi-arabia/

Cobbinah, P. B., Black, R., & Thwaites, R. (2011). Reflections on six decades of the concept of development: Evaluation and future research. *Journal of Sustainable Development in Africa, 13*(7), 134–149.

Cobbinah, P. B., Erdiaw-Kwasie, M. O., & Amoateng, P. (2015). Rethinking sustainable development within the framework of poverty and urbanisation in developing countries. *Environmental Development, 13*, 18–32.

Cobbinah, P. B., & Finn, B. M. (2022). Planning and climate change in African cities: Informal urbanization and "just" urban transformations. *Journal of Planning Literature*, 08854122221128762.

Cohen, B. (2006). Urbanization in developing countries: Current trends, future projections, and key challenges for sustainability. *Technology in Society, 28*(1–2), 63–80.

Conedera, M., Del Biaggio, A., Seeland, K., Moretti, M., & Home, R. (2015). Residents' preferences and use of urban and peri-urban green spaces in a Swiss mountainous region of the Southern Alps. *Urban Forestry & Urban Greening, 14*(1), 139–147.

Côté-Roy, & Moser, S. (2022). "Fast urban model-making": Constructing Moroccan urban expertise through Zenata Eco-City. *Urban Geography* (ahead-of-print), 1–21. https://doi.org/10.1080/02723638.2022.2131261

Croce, S., & Vettorato, D. (2021). Urban surface uses for climate resilient and sustainable cities: A catalogue of solutions. *Sustainable Cities and Society, 75*, Article 103313.

Cube, C. (2021, November 1). Resilience 2021: Building resilience through Bim. National Institute of Building Sciences. Available online https://www.nibs.org/blog/resilience-2021-building-resilience-through-bim

Cutter, S. L., Ahearn, J. A., Amadei, B., Crawford, P., Eide, E. A., Galloway, G. E., & Zoback, M. L. (2013). Disaster resilience: A national imperative. *Environment: Science and Policy for Sustainable Development, 55*(2), 25–29.

Credable. (2022). *NEOM: World's biggest billion dollar smart city plan.* Credable. Available online https://www.credable.in/insights-by-credable/neom-worlds-biggest-billion-dollar-smart-city-plan/. Accessed August 12, 2023.

Dastjerdi, M. S., Lak, A., Ghaffari, A., & Sharifi, A. (2021). A conceptual framework for resilient place assessment based on spatial resilience approach: An integrative review. *Urban Climate, 36*, Article 100794.

El Mallah, F. (2019). Chapter 1: Climate change and the challenges facing Arab Cities. In *Local climate action in the Arab region: Lessons learned and way forward.* UN-Habitat. Available on https://www.climamed.eu/wp-content/uploads/files/UN-Habitat_Local_Climate_Action_Final_digital.pdf

Ewing, L., & Synolakis, C. (2011). Coastal resilience: Can we get beyond planning the last disaster? In *Solutions to Coastal Disasters 2011* (pp. 936–947).

Fang, L., & Lerner, S. (2019, September 18). *Saudi Arabia denies its key role in climate change even as it prepares for the worst.* The Intercept. Available online https://theintercept.com/2019/09/18/saudi-arabia-aramco-oil-climate-change/

Gazzeh, K., & Abubakar, I. R. (2018). Regional disparity in access to basic public services in Saudi Arabia: A sustainability challenge. *Utilities Policy, 52*, 70–80.

General Directorate of Civil Defence (GDCD). (2020). *National plan for natural disasters risk reduction.* Ministry of Interior. Available online https://www.998.gov.sa/Ar/CivilDefence

Govindarajulu, D. (2014). Urban green space planning for climate adaptation in Indian cities. *Urban Climate, 10*, 35–41.

G20 Climate Risk Atlas. (2021, October 28). *G20 climate risk atlas Saudi Arabia.* https://www.g20climaterisks.ouwaiudi-arabia/

Hameed, A., Jabeen, I., & Afzal, N. (2022). Towards an eco-friendly future: A corpus-based analysis of media discourse on the Saudi Green Initiative. *Lege Artis, 7*.

Hamouda, S., & Okumus, O. (2019). Chapter 5: Access to climate finance and project implementation. In*Local climate action in the Arab region: Lessons learned and way forward.* UN-Habitat. Available online https://unhabitat.org/sites/default/files/2019/10/un-habitat_local_climate_action_final_digital_high_res_compressed.pdf

Hosseini, S., Barker, K., & Ramirez-Marquez, J. E. (2016). A review of definitions and measures of system resilience. *Reliability Engineering & System Safety, 145*, 47–61.

Infield, E. M. H., Abunnasr, Y., & Ryan, R. L. (Eds.). (2018). *Planning for climate change: A reader in green infrastructure and sustainable design for resilient cities.* Routledge.

Jaafar, B., Said, I., Reba, M. N. M., & Rasidi, M. H. (2013). Impact of vertical greenery system on internal building corridors in the tropic. *Procedia-Social and Behavioral Sciences, 105*, 558–568.

Jim, C. Y. (2013). Sustainable urban greening strategies for compact cities in developing and developed economies. *Urban Ecosystems, 16*, 741–761.

Kane, F. (2021). Saudi Arabia's methane pledge is a big step forward in fight against climate change. *Arab News*. Available online https://www.arabnews.com/node/1954421

Khan, M. U. H. (2016). Saudi Arabia's vision 2030. *Defence Journal, 19*(11), 36.

Kojima, K., Fujita, K., & Takewaki, I. (2014). Building earthquake resilience in sustainable cities in terms of input energy. *Sustainable Cities and Society, 12*, 46–62.

Koks, E. E., Rozenberg, J., Zorn, C., Tariverdi, M., Vousdoukas, M., Fraser, S. A., & Hallegatte, S. (2019). A global multi-hazard risk analysis of road and railway infrastructure assets. *Nature Communications, 10*(1), 2677.

Kuo, N. W., Sharifi, A., & Li, C. E. (2022). Smart cities and urban resilience: Insights from a Delphi survey. *Resilient smart cities: Theoretical and empirical insights* (pp. 119–138). Springer International Publishing.

Lam, N. S., Reams, M., Li, K., Li, C., & Mata, L. P. (2016). Measuring community resilience to coastal hazards along the Northern Gulf of Mexico. *Natural Hazards Review, 17*(1), 04015013.

Ledraa, T. A., & Al-Ghamdi, A. M. (2020). Planning and management issues and challenges of flash flooding disasters in Saudi Arabia: The case of Riyadh City. *Journal of Architecture and Planning, 32*, 155–171.

Leichenko, R. (2011). Climate change and urban resilience. *Current Opinion in Environmental Sustainability, 3*(3), 164–168.

Liao, K. H. (2012). A theory on urban resilience to floods—A basis for alternative planning practices. *Ecology and Society, 17*(4).

MacDonald, F. (2022). *Kuwait discusses steps to lower emissions as temperatures rise.* Bloomberg. Available online https://www.bloomberg.com/news/articles/2022-01-18/kuwait-discusses-steps-to-lower-emissions-as-temperatures-rise?utm_content=business&utm_campaign=socialflow-organic&utm_source=twitter&cmpid=socialflow-twitter-business&utm_medium=social#xj4y7vzkg

Mahajan, S., Hausladen, C. I., Sánchez-Vaquerizo, J. A., Korecki, M., & Helbing, D. (2022). Participatory resilience: Surviving, recovering and improving together. *Sustainable Cities and Society, 83*, Article 103942.

Mahdaoui, K., & Alibou, J. (2021). Urban resilience in a context of climate change. In *E3S Web of Conferences, 234*, 00053. EDP Sciences.

Mahmood, H., Tawfik, T., & Maham Furqan. (2020). Oil sector and CO_2 emissions in Saudi Arabia: Asymmetry analysis. *Palgrave Communications, 6*(1). https://doi.org/10.1057/s41599-020-0470-z

Marana, P., Eden, C., Eriksson, H., Grimes, C., Hernantes, J., Howick, S., & Serrano, N. (2019). Towards a resilience management guideline—Cities as a starting point for societal resilience. *Sustainable Cities and Society, 48*, Article 101531.

Mead, L. (2022). *Disaster risk reduction in an unstable world*. IISD. Available online https://www.iisd.org/articles/disaster-risk-reduction

Meerow, S., & Newell, J. P. (2017). Spatial planning for multifunctional green infrastructure: Growing resilience in Detroit. *Landscape and Urban Planning, 159*, 62–75.

Meerow, S., Newell, J. P., & Stults, M. (2016). Defining urban resilience: A review. *Landscape and Urban Planning, 147*, 38–49.

Ministry of Municipal and Rural Affairs. (2015). *Review of the National Spatial Strategy*. Available online https://unhabitat.org/sites/default/files/download-manager-files/FSCP_Draft%20NSS%20Review%20%2820151130%29.pdf

MWAN. (2023). *Waste management sector's objectives for 2035*. Available online https://mwan.gov.sa/en

Nahiduzzaman, K. M., Aldosary, A. S., & Rahman, M. T. (2015). Flood induced vulnerability in strategic plan making process of Riyadh city. *Habitat International, 49*, 375–385.

Newman, P., Beatley, T., & Boyer, H. (2009). Resilient cities: Responsing to peak oil and climate change. *Journal of Urban Design, 17*(2). https://doi.org/10.1080/13574809.2012.666179

Ochoa, J. J., Tan, Y., Qian, Q. K., Shen, L., & Moreno, E. L. (2018). Learning from best practices in sustainable urbanization. *Habitat International, 78*, 83–95.

Papa, R., Galderisi, A., Vigo Majello, M. C., & Saretta, E. (2015). Smart and resilient cities. A systemic approach for developing cross-sectoral strategies in the face of climate change. *TeMA Journal of Land Use, Mobility and Environment, 8*(1), 19–49.

Page, M. J., McKenzie, J. E., Bossuyt, P. M., Boutron, I., Hoffmann, T. C., Mulrow, C. D., & Moher, D. (2021). *The PRISMA 2020 statement: An updated guideline for reporting systematic reviews*. https://www.bmj.com/content/372/bmj.n71

Radwan, R. (2022, November13). How Saudi Arabia is translating its climate-change ambitions into action. *Arab News*. https://www.arabnews.com/node/2198686/business-economy

Radwan, R. (2023, March 29). How the Saudi Green Initiative has moved from ambition to action, two years on. *Arab News*. Available online https://www.arabnews.com/node/2277406/saudi-arabia

Rapoport, E. (2014). Utopian visions and real estate dreams: The eco-city past, present and future. *Geography Compass, 8*(2), 137–149.

Salehi, P., & Sharifi, A. (2022). Toward integrating resilience thinking in smart city planning and development. *Resilient smart cities: Theoretical and empirical insights* (pp. 459–466). Springer International Publishing.

Saudi Green Initiative. (2021). *Championing climate action at home and abroad*. Available online https://saudigreeninitiative.org

Sarkis-Onofre, R., Catalá-López, F., Aromataris, E., & Lockwood, C. (2021). *How to properly use the PRISMA Statement*. https://systematicreviewsjournal.biomedcentral.com/articles/10.1186/s13643-021-01671-z

Sharif, M. (2015). Analysis of projected temperature changes over Saudi Arabia in the twenty-first century. *Arabian Journal of Geosciences, 8*, 8795–8809.

Shabaneh, R., Rioux, B., & Haji, F. B. (2022). *Moving Saudi Arabia's role in the global methane pledge forward*. KAPSARC. Available online https://www.kapsarc.org/research/publications/moving-saudi-arabias-role-in-the-global-methane-pledge-forward/

Sharifi, A. (2019). Urban form resilience: A meso-scale analysis. *Cities, 93*, 238–252.

Sharifi, A. (2020). Urban resilience assessment: Mapping knowledge structure and trends. *Sustainability, 12*(15), 5918.

Sharifi, A. (2022a). Sustainability and resilience co-benefits and trade-offs of urban climate change adaptation and mitigation measures. *Handbook of climate change mitigation and adaptation* (pp. 1369–1403). Springer International Publishing.

Sharifi, A. (2022b). Smart city indicators: Towards exploring potential linkages to disaster resilience abilities. *APN, 12*(1), 76–90.

Sharifi, A., & Allam, Z. (2022). On the taxonomy of smart city indicators and their alignment with sustainability and resilience. *Environment and Planning B: Urban Analytics and City Science, 49*(5), 1536–1555.

Sharifi, A., Khavarian-Garmsir, A. R., & Kummitha, R. K. R. (2021). Contributions of smart city solutions and technologies to resilience against the COVID-19 pandemic: A literature review. *Sustainability, 13*(14), 8018.

Sharifi, A., Srivastava, R., Singh, N., Tomar, R., & Raji, M. A. (2022). Recent advances in smart cities and urban resilience and the need for resilient smart cities. *Resilient smart cities: Theoretical and empirical insights*, 17–37.

Sharifi, A., & Yamagata, Y. (2016). Principles and criteria for assessing urban energy resilience: A literature review. *Renewable and Sustainable Energy Reviews, 60*, 1654–1677.

Shehri, T. A., Braun, J. F., Howarth, N., Lanza, A., & Luomi, M. (2022). Saudi Arabia's climate change policy and the circular carbon economy approach. *Climate Policy*, 1–17.

Sileyew, K. J. (2019). *Research design and methodology* (pp. 1–12). IntechOpen.

Sturiale, L., & Scuderi, A. (2019). The role of green infrastructures in urban planning for climate change adaptation. *Climate, 7*(10), 119.

Tambo, E., Fouad, A. M., & Khater, E. I. (2017). Strengthening community emergency preparedness and response in threats and epidemics disasters prevention and management in Saudi Arabia. *International Journal of Emergency Management, 13*(3), 288–303.

UAE Government Portal. (2022). *The UAE's response to climate change*. Available online https://u.ae/en/information-and-services/environment-and-energy/climate-change/theuaesresponsetoclimatechange#:~:text=In%20order%20to%20control%20the,and%20undertaking%20many%20other%20steps

URB (2022, October 24). *Alnama Smart City. A zero carbon smart city where people, planet and profit thrive in balance*. Available online https://urb.ae/projects/alnamasmartcity/

UN-Habitat. (2016). *Country profile Saudi Arabia*. Available online https://unhabitat.org/sites/default/files/download-manager-files/Country%20Profile%20Saudi%20Arabia%20English.pdf

UN-Habitat. (2019). *Saudi Cities Report 2019*. Available online https://unhabitat.org/sites/default/files/2020/05/saudi_city_report.english.pdf

UN-Habitat. (2022a). *Smart sustainable cities and smart digital solutions for urban resilience in the Arab Region: Lessons from the pandemic*. Available online https://www.unescwa.org/publications/smart-cities-digital-urban-arab-region

UN-Habitat. (2022b). "Climate change strategy for the Arab region 2022 - 2025." Available online https://unhabitat.org/sites/default/files/2022/10/ccs_-_roas_-_final_-_a4_sheets.pdf

United Nations Climate Change (UNCC). (2023). *Key takeaways from COP 27*. Available online https://unfccc.int/

United Nations Office for Disaster Risk Reduction (UNDRR). (2023). *Sendai framework*. Available online https://www.preventionweb.net/sendai-framework/united-nations-plan-action-drr-resilience#:~:text=The%20UN%20Plan%20of%20Action,2030%20Agenda%20for%20Sustainable%20Development

USGBC. (2023). *Green building leadership is LEED*. Available online https://new.usgbc.org/leed

Van Krieken, T., Kulatunga, U., & Pathirage, C. (2017, September). Importance of community participation in disaster recovery. In *13th IPGRC 2017 Full Conference Proceedings* (pp. 860–869). University of Salford.

Vaughn, R. M., Hostetler, M., Escobedo, F. J., & Jones, P. (2014). The influence of subdivision design and conservation of open space on carbon storage and sequestration. *Landscape and Urban Planning, 131*, 64–73.

Vishwa, G. (2021). *Saudi Arabia smart cities market by functional area*. Allied Market Research. Available online https://www.alliedmarketresearch.com/saudi-arabia-smart-cities-market-A10247

Vourvachis, P., & Woodward, T. (2015). Content analysis in social and environmental reporting research: Trends and challenges. *Journal of Applied Accounting Research, 16*(2), 166–195.

Wamsler, C., Brink, E., & Rivera, C. (2013). Planning for climate change in urban areas: From theory to practice. *Journal of Cleaner Production, 50*, 68–81.

Wang, L. (2022). Exploring a knowledge map for urban resilience to climate change. *Cities, 131*, Article 104048.

Wang, J., Cao, S.-J., & Yu, C. W. (2020). Development trend and challenges of sustainable urban design in the digital age. *Indoor and Built Environment, 30*(1), 3–6. https://doi.org/10.1177/1420326x20976058

Wogan, D., Carey, E., & Cooke, D. (2019). *Policy pathways to meet Saudi Arabia's contribution to the Paris Agreement*. KAPSARC. Available online https://doi.org/10.30573/ks--2019-dp49
Zhou, Q., Zhu, M., Qiao, Y., Zhang, X., & Chen, J. (2021). Achieving resilience through smart cities? Evidence from China. *Habitat International, 111*, Article 102348.

Chapter 6
Green Infrastructure: What It Means for Climate Adaptation and Planning in Saudi Arabia

6.1 Introduction: Green Infrastructure and Urban Resilience in Arid Contexts

As discussed in previous chapters, urbanization in arid and semi-arid regions has emerged as one of the most pressing and complex planning challenges of the twenty-first century. Unlike urban growth in more climatically moderate or water-abundant environments, urbanization in desert contexts unfolds under particularly severe climatic and ecological constraints that fundamentally shape how cities function, expand, and adapt. Extreme heat, chronic water scarcity, fragile ecosystems, and limited natural vegetation impose limits on conventional urban development models and require planners to rethink the relationship between urban form, infrastructure, and environmental resilience. As urban populations in arid regions continue to grow, these constraints increasingly shape not only environmental outcomes, but also dimensions of social equity, public health, and the long-term viability of urban systems (Almulhim & Cobbinah, 2023; Meerow et al., 2019; Siddiqui et al., 2025).

Saudi Arabia exemplifies these challenges with particular intensity (see Chapters 2, 3, and 4). Rapid urban expansion over recent decades has transformed Saudi cities into major metropolitan centers operating almost entirely within arid and hyper-arid climatic zones. High summer temperatures, recurrent heatwaves, minimal precipitation, and extreme evapotranspiration rates combine to produce persistent conditions of thermal stress and water insecurity. These climatic pressures are not episodic anomalies but structural characteristics of the Saudi urban environment, shaping everyday urban life as well as long-term planning and policy decisions. In this context, Saudi cities are simultaneously highly vulnerable to climate impacts and central to national efforts aimed at climate adaptation and urban resilience (Almulhim & Cobbinah, 2024a; Almulhim & Cobbinah, 2025).

Within contemporary urban planning discourse, green infrastructure (GI) has gained increasing prominence as a response to a range of environmental and climate-related urban challenges. It is commonly conceptualized as a strategically planned

A. I. Almulhim and P. B. Cobbinah, *Urbanization Rethought*, Urban Sustainability,
https://doi.org/10.1007/978-981-92-0118-1_6

network of natural and semi-natural areas designed to deliver multiple ecosystem services across different spatial scales, extending beyond isolated green spaces or conventional landscaping practices. Rather than functioning as a single-purpose environmental intervention, GI is widely framed as a multifunctional planning system capable of integrating ecological processes, social benefits, and climatic functions within urban development frameworks. This multifunctional framing has positioned GI at the core of resilience-oriented planning agendas, particularly within broader debates on climate adaptation, sustainability, and urban liveability. In this context, GI is increasingly understood not as an aesthetic or supplementary urban feature, but as a form of infrastructure whose performance depends on spatial planning, governance arrangements, and long-term integration with other urban systems (Lafortezza et al., 2013; Monteiro et al., 2020).

Despite the growing prominence of GI within global urban planning discourse, its transferability to arid and water-scarce urban environments remains deeply contested. Much of the conceptual and operational foundation of contemporary GI approaches has been developed in temperate contexts, where climatic conditions, water availability, and vegetation dynamics differ fundamentally from those of desert cities. When such models are applied uncritically in arid contexts, they risk overlooking structural environmental constraints related to water scarcity, ecological suitability, and long-term maintenance demands, thereby limiting their effectiveness as resilience-oriented interventions (Moscatelli & Raffa, 2023). Recent studies in arid and semi-arid urban contexts indicate that GI initiatives may, under certain conditions, shift from being resilience-enhancing strategies to resource-intensive systems that exacerbate existing urban vulnerabilities (Okour & Shaweesh, 2024). High irrigation requirements, dependence on non-renewable or energy-intensive water sources, fragmented governance arrangements, and limited institutional capacity have been identified as key barriers that constrain the sustainable planning and implementation of GI in water-scarce environments (Okour & Shaweesh, 2024). These challenges underscore the need to move beyond universalized or normative interpretations of GI and instead adopt context-sensitive planning frameworks that explicitly foreground environmental limits, resource efficiency, and long-term urban resilience in arid settings (Moscatelli & Raffa, 2023; Okour & Shaweesh, 2024).

Saudi Arabia occupies a critical position in contemporary debates on GI and urban resilience in arid contexts. On the one hand, national climate and development frameworks increasingly emphasize climate adaptation, environmental sustainability, and resilience building, explicitly recognizing aridity, water scarcity, desertification, and extreme heat as defining development constraints, while framing urban areas as central arenas for addressing climate-related risks through integrated planning and infrastructure interventions (Almulhim & Cobbinah, 2024a; Government of Saudi Arabia, 2025). This policy recognition elevates the role of urban planning and infrastructure systems in responding to climate stressors and situates GI within broader national adaptation narratives. On the other hand, the scale and severity of environmental constraints facing Saudi cities raise fundamental questions regarding the feasibility and effectiveness of prevailing GI approaches when applied under conditions of extreme aridity and chronic water scarcity (Moscatelli & Raffa, 2023).

Rapid urbanization, high resource consumption, and institutional and governance challenges place additional pressures on urban systems, complicating efforts to align resilience-oriented planning ambitions with environmental realities. This juxtaposition positions Saudi Arabia as a critical case for rethinking GI in arid urban planning, as Saudi cities function as stress tests for resilience-oriented planning concepts that reveal both the aspirations embedded in national policy frameworks and the limits imposed by biophysical and institutional constraints (Almulhim & Cobbinah, 2024a). Examining GI within this context therefore offers insights not only for Saudi Arabia, but also for other arid and resource-constrained urban regions confronting similar challenges.

This chapter argues that GI should not be treated as a universally applicable or inherently sustainable planning solution. Instead, its role in arid cities must be evaluated through a planning-centered perspective that foregrounds climate adaptation, resource efficiency, and long-term resilience. GI is conceptualized here not as an aesthetic enhancement or symbolic environmental gesture, but as a component of urban infrastructure systems whose performance depends on land-use planning, urban form, governance capacity, and environmental context. In arid environments, the effectiveness of GI is inseparable from questions of water management, institutional coordination, and planning integration.

The objectives of this chapter are threefold. First, it reconsiders dominant interpretations of GI and clarifies its distinction from conventional urban greening practices, particularly in arid and water-scarce contexts. Second, it examines the specific climate adaptation imperatives that shape planning decisions in desert cities, with particular attention to heat stress, water scarcity, and environmental limits. Third, drawing on the Saudi urban experience, the chapter explores how GI can be planned in ways that enhance urban resilience without reproducing unsustainable resource dependencies.

The chapter is structured as follows. Section 6.2 revisits the concept of GI and critically distinguishes it from landscaping and beautification approaches, emphasizing its role as a planning system rather than a collection of isolated green spaces. Section 6.3 outlines the climate adaptation imperatives that define arid urban environments, highlighting the constraints that shape planning choices in Saudi cities. Section 6.4 examines the challenges of planning GI under conditions of water scarcity, focusing on trade-offs, governance constraints, and alternative planning strategies. Section 6.5 situates Saudi cities as extreme test cases for GI implementation, drawing broader lessons from their experience. Section 6.6 discusses the implications of these findings for urban resilience and planning practice, before Section 6.7 concludes by outlining principles for context-sensitive GI planning in arid urban environments.

By situating GI within the realities of arid urbanization and climate adaptation, this chapter contributes to the broader aim of the book to rethink urbanization, planning, and resilience in Saudi Arabia, while offering insights applicable to desert cities globally.

6.2 Rethinking Green Infrastructure Beyond Urban Greening

GI has become a widely used concept in contemporary urban planning, frequently invoked as a response to a broad range of environmental and climatic challenges facing cities (Monteiro et al., 2020). Within both policy discourse and planning practice, GI is often associated with urban greening initiatives such as parks, tree planting, green corridors, and landscaped public spaces. While such elements are not inherently problematic, the tendency to equate GI with visible greenery risks oversimplifying a concept that was originally developed to address systemic relationships between urban form, ecological processes, and infrastructure networks (Benedict & MacMahon, 2002; Mell & Clement, 2019). From a planning perspective, GI was initially conceptualized as a strategic and interconnected system designed to perform functions traditionally associated with gray infrastructure, while simultaneously delivering ecological and social benefits (Lafortezza et al., 2013). Rather than operating as a collection of isolated green spaces, GI was intended to function as a networked framework integrated into land-use planning, transportation systems, water management, and broader urban development patterns (Benedict & MacMahon, 2002; Monteiro et al., 2020). This systemic understanding underpins the relevance of GI for urban resilience, particularly in contexts where cities face increasing climatic and environmental stress and highlights the need to move beyond narrow greening interpretations toward more comprehensive planning-oriented conceptualizations of GI (Lafortezza et al., 2013; Mell & Clement, 2019).

However, as the concept gained popularity, its application in many cities shifted toward more symbolic and aesthetic interpretations. GI increasingly became synonymous with beautification, place-making, and environmental branding, often detached from broader planning logics (Mell & Clement, 2019). This shift has important implications for arid urban environments, where resource constraints demand careful scrutiny of how and why GI is implemented. When GI is reduced to landscaping, it risks undermining the very resilience objectives it is meant to support. In arid cities, the distinction between urban greening and GI is not merely semantic but fundamentally planning-related. Urban greening typically emphasizes visual enhancement, recreational value, and short-term environmental improvements, often without systematic consideration of water use, maintenance requirements, or long-term sustainability. GI, by contrast, must be understood as an infrastructural system whose performance depends on integration with urban planning instruments, environmental constraints, and governance structures. This distinction is particularly critical in water-scarce environments, where poorly planned greening initiatives can exacerbate resource pressures rather than alleviate them.

Reframing GI beyond greening requires a return to its infrastructural logic. In arid contexts, GI should be evaluated in terms of its functional contribution to urban resilience, including its capacity to moderate urban temperatures, manage environmental stress, and support adaptive urban forms. This evaluation must move beyond assumptions that more greenery automatically translates into greater resilience. Instead, planners must assess how different forms of GI interact with urban morphology, density, land use, and climatic conditions.

The Saudi urban context illustrates the consequences of conflating GI with greening. Ambitious narratives around environmental improvement and urban liveability have elevated the visibility of green spaces within urban development discourse (Government of Saudi Arabia, 2025; Almulhim & Cobbinah, 2024a). Yet the climatic reality of extreme heat, minimal rainfall, and high evapotranspiration rates challenges the viability of vegetation-intensive approaches. Without careful planning, such approaches risk creating landscapes that are heavily dependent on artificial irrigation, energy-intensive water production, and continuous maintenance, conditions that conflict with principles of long-term resilience.

From a planning standpoint, GI in arid cities must therefore be understood as a context-sensitive system, not a universal template. This requires recognizing that GI does not always take the form of lush vegetation or expansive green spaces. In desert environments, effective GI may prioritize shaded corridors, compact urban form, permeable surfaces, and climate-responsive design that reduces exposure to heat while minimizing water demand. These strategies may appear less visually "green" but can be more resilient and sustainable over time.

Importantly, rethinking GI also involves reconsidering its relationship with urban form. Many greening initiatives focus on adding green elements to existing urban layouts without questioning whether those layouts themselves contribute to climate vulnerability. Low-density development, wide road networks, and fragmented land-use patterns can intensify heat exposure and increase the spatial demand for infrastructure (Almulhim & Cobbinah, 2024b). In such contexts, GI interventions risk becoming compensatory measures that attempt to mitigate problems created by underlying planning decisions. A planning-centered approach to GI instead emphasizes alignment between GI strategies and broader urban form objectives, such as compactness, mixed land use, and reduced surface exposure.

Governance and institutional capacity further shape how GI is interpreted and implemented. When GI is treated as a design or beautification issue, responsibility often falls to municipal departments focused on landscaping or public amenities, rather than planning authorities responsible for land use and infrastructure coordination (Mell & Clement, 2019). This institutional fragmentation can limit the effectiveness of GI strategies, particularly in arid environments where coordination between water management, urban planning, and environmental regulation is essential. Rethinking GI beyond greening thus also requires repositioning it within planning institutions and decision-making processes.

Another critical dimension is the temporal scale at which GI is evaluated. Greening initiatives often prioritize short-term visibility and immediate benefits, while overlooking long-term performance under climatic stress. In arid cities, the sustainability of GI depends on its ability to function under prolonged heat, water scarcity, and maintenance constraints over time. Planning for GI must therefore incorporate lifecycle considerations, including resource inputs, institutional capacity for maintenance, and adaptability to changing climatic conditions. Without this temporal perspective, GI risks becoming a liability rather than an asset.

Reconceptualizing GI in this way does not imply rejecting it as a planning tool. Rather, it underscores the need to deploy GI selectively and strategically within arid urban environments. When grounded in planning logic, GI can contribute to urban resilience by supporting microclimate regulation, enhancing public space usability, and integrating ecological considerations into urban development. However, these contributions are contingent on moving beyond generic greening models toward approaches that are responsive to environmental limits and planning realities.

In the Saudi context, this reconceptualization is particularly urgent. As cities continue to expand and adapt to climatic pressures, the choices made around GI will shape not only environmental outcomes but also institutional practices and public expectations. Treating GI as an infrastructural system, rather than an aesthetic add-on, creates space for more honest and effective planning debates about what forms of urban resilience are feasible in arid environments.

In summary, rethinking GI beyond urban greening is a necessary step toward aligning resilience ambitions with environmental realities in arid cities. For planners, this means shifting attention from the quantity and visibility of green spaces to the quality, functionality, and integration of GI within broader urban systems. This reframing sets the foundation for examining the specific climate adaptation imperatives of arid urban environments, which are explored in the following section.

6.3 Climate Adaptation Imperatives in Arid Urban Environments

Climate adaptation has become a central concern in urban planning as cities confront increasing exposure to climate-related stresses. In arid and semi-arid regions, however, adaptation imperatives differ fundamentally from those in temperate or water-abundant environments. Rather than responding to newly emerging climatic risks, cities in desert contexts operate under long-standing conditions of extreme heat, water scarcity, and environmental variability that shape urban life on a daily basis. In such settings, climate adaptation is not an abstract future challenge but a structural planning concern embedded in the functioning of urban systems (Almulhim & Cobbinah, 2024a; Meerow et al., 2019).

A defining adaptation challenge in arid cities is urban heat stress and thermal discomfort. Extremely high ambient temperatures, prolonged heatwaves, and elevated night-time temperatures generate persistent thermal stress that affects public health, energy demand, and the usability of urban space. Dense built environments, extensive paved surfaces, and limited shading intensify heat accumulation, producing severe urban heat island effects. For planners, this translates into challenges related to outdoor mobility, public space design, walkability, and social interaction, particularly during extended periods of extreme heat. Addressing thermal discomfort is therefore a central adaptation priority that directly influences urban form and planning decisions (Almulhim & Cobbinah, 2023; Liu et al., 2025a).

Water scarcity constitutes a second, and equally critical, constraint on climate adaptation in arid urban environments. Limited natural water availability, high evaporation rates, and minimal rainfall fundamentally shape the feasibility of adaptation strategies. Unlike cities in humid regions, where water can support a wide range of climate-responsive interventions, arid cities must plan adaptation measures under strict resource limitations. From a planning perspective, water scarcity affects land-use patterns, urban density, infrastructure provision, and the long-term sustainability of environmental interventions. Any adaptation strategy, particularly those involving GI, must therefore be evaluated against its water demand and long-term resource implications (Almulhim & Cobbinah, 2024a; Almulhim & Cobbinah, 2023).

Beyond heat and water scarcity, arid cities are also exposed to climate extremes that affect urban liveability in less visible but equally significant ways. Dust storms reduce air quality and limit the usability of outdoor environments, while prolonged drought conditions exacerbate environmental stress and place additional pressure on urban services. At the same time, infrequent but intense rainfall events can overwhelm drainage systems and disrupt urban functioning. These climatic extremes challenge conventional assumptions about resilience and require planners to account for a wider range of environmental stressors when designing urban systems (Almulhim & Cobbinah, 2023).

Together, these conditions highlight the centrality of adaptation priorities over mitigation ambitions in arid urban contexts. While climate mitigation remains an important long-term objective, the immediacy and severity of heat stress, water scarcity, and environmental extremes make adaptation the dominant concern in planning practice. In desert cities, reducing exposure to climate stress and maintaining basic urban functionality often take precedence over emissions reduction strategies that do not directly address lived urban vulnerability. This prioritization does not diminish the importance of mitigation but reflects the practical realities faced by planners operating under extreme climatic constraints.

The Saudi Arabian urban context illustrates these adaptation imperatives with particular clarity. Cities across Saudi Arabia experience some of the most extreme combinations of heat and aridity globally, making climate adaptation a structural planning concern rather than a marginal policy issue. These pressures are not hypothetical; they are explicitly acknowledged in national climate policy, which recognizes aridity, water scarcity, and extreme heat as defining development constraints and positions cities as key sites for climate adaptation. This recognition reinforces the need to ground urban planning decisions in adaptation priorities that reflect environmental realities (Government of Saudi Arabia, 2025).

For planning practice in Saudi cities, these adaptation imperatives carry important implications. First, planners operating in desert environments must move beyond generic climate-responsive models and adopt strategies that explicitly account for extreme heat, water scarcity, and arid urban morphology. Second, climate adaptation needs to be embedded within core planning instruments, such as land-use regulation, density controls, and spatial structure, rather than treated as a stand-alone environmental objective. Third, planning interventions that are often framed as environmentally beneficial in other contexts require critical reassessment in Saudi cities, particularly in terms of their water intensity, institutional feasibility, and long-term operational viability under arid conditions.

These considerations are particularly relevant for evaluating the role of GI in Saudi cities. While GI is often promoted as a climate adaptation tool, its effectiveness depends on how well it responds to the specific imperatives of heat stress, water scarcity, and environmental extremes outlined in this section. Without careful planning, GI risks becoming misaligned with adaptation priorities and undermining long-term resilience.

In summary, climate adaptation imperatives in arid urban environments are defined by persistent heat stress, chronic water scarcity, and exposure to environmental extremes that directly affect urban liveability. These conditions prioritize adaptation over mitigation in planning practice and demand context-sensitive strategies that align environmental ambition with resource constraints. Understanding these imperatives is essential for assessing whether and how GI can contribute meaningfully to urban resilience in arid cities. The following section examines this question by exploring the planning of GI under conditions of water scarcity.

6.4 Planning Green Infrastructure Under Water Scarcity

Planning GI in arid urban environments is fundamentally shaped by water scarcity. While GI is frequently promoted as a climate adaptation strategy capable of enhancing urban resilience, its feasibility and long-term sustainability in desert cities depend on how water constraints are accounted for in planning decisions and long-term management practices. In water-abundant contexts, the expansion of GI can proceed with relatively limited concern for resource availability. In arid cities, by contrast, water scarcity operates as a binding constraint that reshapes the scale, form, and performance of GI interventions (Okour & Shaweesh, 2024).

From a planning perspective, GI in arid cities must be understood within a broader water–energy–urban planning nexus. Water used to support urban environments is not a passive resource, but is embedded in energy-intensive supply and management systems. As a result, planning decisions related to land use, density, and infrastructure coordination influence not only urban form, but also energy demand and resource efficiency. GI therefore cannot be treated as an isolated environmental intervention, but must be approached as a planning instrument embedded within interconnected urban systems (Ruan et al., 2023).

National climate and adaptation policy in Saudi Arabia explicitly acknowledges the importance of integrated water management as a core component of addressing climate vulnerability. Policy frameworks emphasize improving water efficiency, expanding the reuse of treated wastewater, and reducing pressure on scarce freshwater resources as central adaptation priorities in an arid national context. Within this policy framing, urban areas are positioned as key sites for implementing integrated water and climate strategies, reflecting an ambition to align adaptation objectives with urban development processes and infrastructure systems. This recognition implicitly situates GI within a broader water–energy–urban planning nexus and signals a policy-level commitment to integrated resource management rather than isolated environmental interventions (Government of Saudi Arabia, 2025).

However, the existence of policy ambition does not, in itself, resolve the planning challenges associated with implementing GI under conditions of water scarcity. The critical question for urban planners is not whether integrated water management is endorsed at the national level, but how such ambitions translate into concrete decisions related to urban form, land-use regulation, zoning, and the spatial configuration of GI. In arid cities, reliance on treated wastewater to support GI raises complex questions regarding infrastructure capacity, spatial distribution, long-term maintenance responsibilities, and coordination between planning and water authorities. Within this broader water–energy–urban planning nexus, GI represents one planning instrument whose performance is directly shaped by water availability and energy intensity.

One of the most significant challenges in this context involves the trade-off between cooling benefits and water consumption. Vegetation-based GI can contribute to microclimate regulation and thermal comfort in hot cities; however, in arid urban environments these benefits are strongly dependent on sustained irrigation under conditions of extreme heat and high evaporation. Evidence from desert cities such as Riyadh demonstrates that, without irrigation, increased GI may fail to deliver cooling benefits, while extensive irrigation substantially increases water consumption and direct soil evaporation, reducing overall efficiency. Planners must therefore assess whether the cooling gains achieved through GI justify the associated water demand over time, particularly when cumulative impacts are considered at the city scale (Gao et al., 2024).

Maintenance, irrigation, and long-term sustainability present further planning challenges. GI in desert environments requires continuous institutional capacity to manage water supply, monitor vegetation health, and adapt maintenance regimes over time. Interventions that appear viable during early implementation phases may become unsustainable as maintenance costs increase, water availability fluctuates, or institutional responsibilities remain fragmented. Planning for GI must therefore incorporate lifecycle considerations, ensuring that maintenance obligations, governance arrangements, and financial resources are aligned with long-term resilience objectives (Gao et al., 2024; Okour & Shaweesh, 2024).

These constraints underscore the importance of alternative adaptation strategies that reduce dependence on water-intensive greening. In arid urban environments, approaches such as xeriscaping and the use of native or drought-tolerant vegetation can significantly lower water demand while maintaining ecological and microclimatic benefits. Similarly, shaded public spaces, arcades, and pedestrian networks can enhance thermal comfort primarily through solar control by reducing direct solar exposure rather than relying solely on vegetation (Radonic et al., 2025; Amores et al., 2023). Collectively, these strategies broaden the understanding of climate-responsive urbanism beyond conventional notions of greening.

Urban morphology also plays a decisive role in shaping adaptation outcomes. Compact urban form, appropriate density, and passive cooling strategies can reduce heat exposure and energy demand more effectively than dispersed development patterns supplemented by GI (Liu et al., 2025b). From a planning perspective, this highlights the limitations of treating GI as a compensatory solution for climate vulnerability created by underlying urban form decisions. Instead, GI should be planned in conjunction with land-use strategies that minimize thermal stress and resource consumption from the outset.

The role of land-use planning, zoning, and density regulation is therefore central to effective GI planning under water scarcity. Decisions regarding where GI is located, how it is scaled, and which urban areas are prioritized must align with broader planning objectives related to land efficiency, mobility, and infrastructure coordination. In arid cities, selective and targeted deployment of GI, focused on high-impact locations such as dense neighborhoods or critical public spaces, is often more resilient than uniform greening strategies applied across the urban fabric (Liu et al., 2025b).

Institutional and governance constraints further complicate GI planning under conditions of water scarcity in Saudi cities. Responsibilities for urban planning, water management, landscaping, and environmental regulation are distributed across multiple agencies, often with limited coordination, reflecting a sectoral approach to urban development that constrains integrated decision-making (Almulhim & Abubakar, 2023). When GI is framed primarily as an environmental or aesthetic initiative, it may fall outside the core mandate of planning authorities, weakening its alignment with land-use decisions and infrastructure investment. In water-scarce environments, such institutional fragmentation can undermine the effectiveness of GI strategies and exacerbate long-term maintenance challenges, as widely documented in the GI governance literature (Mell & Clement, 2019).

The Saudi experience illustrates the risks of expanding GI without sufficiently embedding it within planning systems that account for water scarcity, energy intensity, and institutional capacity. While national policy frameworks articulate ambitious goals for integrated water management and climate adaptation, the translation of these ambitions into urban planning practice remains uneven. This gap highlights a critical vulnerability: without alignment between policy objectives and planning instruments, GI risks imposing long-term resource burdens rather than enhancing urban resilience (Government of Saudi Arabia, 2025; Almulhim & Abubakar, 2023).

At the same time, water scarcity does not preclude a meaningful role for GI in arid cities. Rather, it demands a selective, planning-led approach that recognizes GI as one component within a broader adaptation toolkit. When integrated with land-use planning, urban form, and governance structures, GI can contribute to thermal comfort, public space quality, and environmental performance without undermining resource sustainability.

In summary, planning GI under water scarcity requires confronting fundamental trade-offs between environmental ambition and resource constraints. The water–energy–urban planning nexus, long-term maintenance demands, and governance capacity all shape the feasibility of GI interventions in arid cities. The Saudi context demonstrates both the presence of policy ambition and the planning challenges involved in translating that ambition into resilient urban outcomes. Understanding these dynamics is essential for assessing whether GI can meaningfully support urban resilience in desert environments.

6.5 Saudi Cities as Stress Tests for Green Infrastructure Strategies

Saudi cities represent some of the most extreme test cases for the implementation of GI in contemporary urban planning. Operating within arid and hyper-arid climatic zones, these cities face a combination of environmental, resource, and institutional constraints that push GI strategies to their practical limits. Unlike contexts where GI can be incrementally adjusted or supported by relatively abundant natural resources, Saudi cities confront conditions in which the feasibility, sustainability, and resilience contribution of GI are continuously tested. This makes them analytically valuable not as exemplary models, but as stress tests that expose the strengths and limitations of GI under extreme conditions (Almulhim & Cobbinah, 2024a).

A defining feature of Saudi cities is the scale and intensity of environmental constraints shaping urban development. Extreme heat, chronic water scarcity, high evapotranspiration rates, and limited natural vegetation form the baseline within which planning decisions are made. These conditions magnify the resource implications of GI interventions and reduce the margin for error in planning and implementation. Strategies that may be environmentally benign or resource-efficient in temperate climates can become highly resource-intensive when transplanted into desert environments. As a result, Saudi cities provide a critical context for interrogating whether prevailing GI models can withstand extreme climatic stress without undermining long-term resilience.

At the same time, Saudi Arabia is characterized by high levels of policy ambition in relation to climate adaptation, environmental sustainability, and urban transformation. National climate policy explicitly recognizes aridity, water scarcity, and extreme heat as defining development constraints and positions cities as key arenas for climate adaptation. Urban planning is framed within broader adaptation pillars that emphasize integrated water management, efficiency, and resilience. This policy recognition situates Saudi cities as environments where GI is not merely an optional enhancement, but part of a wider national narrative on adaptation and sustainability. Yet the coexistence of strong policy ambition and severe environmental limits creates a tension that is central to understanding GI outcomes in Saudi cities (Government of Saudi Arabia, 2025).

This tension is particularly evident in the contrast between ambitions and environmental realities. On the one hand, GI is often associated with improved urban liveability, environmental quality, and climate responsiveness. On the other hand, the environmental realities of Saudi cities, especially water scarcity and energy-intensive urban systems, constrain the extent to which such ambitions can be realized sustainably. GI initiatives must operate within urban systems that rely heavily on engineered water supply, extensive energy inputs, and complex institutional arrangements. This context forces planners to confront difficult trade-offs between symbolic environmental improvements and long-term resource efficiency.

The distinction between existing cities and new urban developments further complicates the implementation of GI in Saudi Arabia. In established cities, GI interventions are often retrofitted into urban fabrics shaped by low-density development, wide road networks, and car-oriented planning. These structural characteristics limit the effectiveness of GI by increasing exposure to heat, expanding infrastructure footprints, and raising maintenance demands. Retrofitting GI in such contexts can become a compensatory exercise that attempts to mitigate vulnerabilities embedded in earlier planning decisions, rather than addressing their root causes.

By contrast, new urban developments offer greater formal flexibility to integrate GI from the outset. In theory, master-planned environments allow planners to align GI with land-use patterns, density, and infrastructure systems more coherently. However, even in new developments, environmental constraints remain binding. The availability of treated wastewater, long-term irrigation requirements, and institutional capacity for maintenance continue to shape outcomes. Moreover, the scale of new developments can amplify resource demands, raising questions about whether integrated planning translates into genuinely resilient urban systems or simply redistributes environmental pressures.

One illustrative example of this new development is King Salman Park in Riyadh, a large-scale urban greening initiative embedded within an already dense metropolitan fabric (see Fig. 6.1). Planned on the former site of Riyadh's airport, the project represents an ambitious attempt to introduce extensive green infrastructure at a metropolitan scale within an arid and heat-intensive environment. Its scale, location, and policy prominence position it as a critical case for examining how green infrastructure strategies operate under conditions of structural water scarcity and extreme thermal stress, and how large-scale greening initiatives negotiate the tension between environmental ambition and ecological constraint.

While the project demonstrates the potential for large-scale environmental reconfiguration within existing cities, it simultaneously raises important questions regarding long-term irrigation requirements, maintenance regimes, and the energy implications of sustaining extensive vegetation in a desert climate. The success of such interventions ultimately depends not only on design ambition, but on institutional capacity, water resource governance, and sustained operational commitment over time.

Across both existing and new urban contexts, planning capacity and implementation gaps play a critical role in shaping GI outcomes. While strategic plans and policy frameworks may articulate broad objectives for adaptation and sustainability, their translation into spatial decisions, zoning regulations, and infrastructure coordination is often uneven. Responsibilities for planning, water management, landscaping, and environmental regulation are frequently distributed across multiple institutions, creating coordination challenges that complicate the implementation of GI and weaken accountability for long-term performance and maintenance (Almulhim & Cobbinah, 2025; Mell & Clement, 2019). Operational challenges further reinforce

Fig. 6.1 King Salman Park within Riyadh's urban structure, illustrating its metropolitan scale and integration within existing transport corridors. *Source* (Dogan, 2026)

these implementation gaps. In arid cities, GI requires sustained management of irrigation systems, vegetation health, and supporting infrastructure under extreme climatic conditions, and where maintenance regimes are under-resourced or institutional responsibilities are unclear, interventions risk rapid degradation. These operational realities underscore the importance of viewing GI not as a one-time planning intervention, but as a long-term infrastructural commitment whose success depends on governance capacity and ongoing management as much as on design quality.

The Saudi experience thus highlights a broader lesson about the limits of policy-driven greening in environmentally constrained contexts. The presence of ambitious national frameworks and adaptation narratives does not automatically translate into resilient urban outcomes. Instead, the effectiveness of GI depends on how well policy aspirations are aligned with planning instruments, institutional capacity, and environmental realities. This misalignment is particularly visible in arid cities, where resource constraints magnify the consequences of poorly integrated planning decisions (Government of Saudi Arabia, 2025).

Importantly, the value of Saudi cities as stress tests lies not in identifying best practices, but in revealing failure points, trade-offs, and institutional tensions that may be less visible in more forgiving environments. By examining where and why

GI struggles under extreme conditions, planners can gain clearer insight into the assumptions embedded in GI models and the conditions required for their success. These insights are highly relevant for other arid and semi-arid regions facing rapid urbanization under climate stress.

For arid cities beyond Saudi Arabia, several lessons emerge from this experience. First, GI should be approached as a selective and context-sensitive planning tool rather than a universal solution. Second, environmental ambition must be matched by realistic assessments of resource availability, maintenance capacity, and institutional coordination. Third, the integration of GI with land-use planning, urban form, and infrastructure systems is essential for avoiding compensatory or symbolic interventions that undermine resilience. Finally, arid cities benefit from reframing resilience not in terms of visible greening, but in terms of thermal comfort, resource efficiency, and long-term urban functionality.

In this sense, Saudi cities function as policy-ambitious but environmentally constrained urban laboratories. They reveal both the potential and the limits of GI as a climate adaptation strategy in desert environments. Understanding these dynamics contributes to a more grounded and critical approach to urban resilience, one that prioritizes planning integration and environmental realism over the uncritical transfer of global models.

This section has positioned Saudi cities as stress tests for GI strategies, emphasizing the interplay between environmental constraints, policy ambition, and planning capacity. The following section builds on these insights by examining the broader planning implications for urban resilience, drawing together the chapter's arguments and situating GI within a wider adaptation toolkit for arid urban environments.

6.6 Green Infrastructure and Urban Resilience: Planning Implications

The preceding sections have examined GI within the environmental, institutional, and planning realities of arid urban environments, using Saudi cities as critical stress tests. This section draws together those insights to address a central question: under what conditions does GI enhance urban resilience in arid contexts, and when does it risk becoming an unsustainable planning burden? Answering this question requires moving beyond generalized advocacy for greening and toward a more selective, planning-led understanding of resilience.

GI can enhance urban resilience in arid cities when it is strategically integrated into planning systems rather than applied as a stand-alone intervention. In such contexts, GI contributes positively to resilience when it supports thermal comfort in high-use public spaces, improves microclimatic conditions in dense urban areas, and complements broader strategies aimed at reducing exposure to heat and environmental stress. These benefits are most evident when GI is carefully targeted, limited in scale, and aligned with land-use planning, urban form, and infrastructure capacity. Under these conditions, GI functions as part of a wider adaptation toolkit rather than as a universal solution.

By contrast, GI risks becoming an unsustainable planning burden when it is deployed expansively without sufficient regard for environmental constraints, resource availability, and long-term maintenance capacity. In arid cities, vegetation-intensive greening strategies that rely on continuous irrigation can place significant pressure on water and energy systems, undermining resilience objectives over time. When GI is treated as a symbolic marker of sustainability, rather than as a performance-based planning tool, it may generate short-term visual or political benefits while producing long-term operational and resource challenges. Such outcomes are particularly problematic in environments where water scarcity magnifies the consequences of poorly integrated planning decisions.

The distinction between resilience-enhancing and burden-inducing GI is therefore not inherent to GI itself, but to how it is planned, governed, and integrated. In arid contexts, resilience depends less on the quantity of GI than on its spatial logic, functional role, and institutional support. This insight challenges dominant narratives that equate resilience with visible greening and underscores the importance of planning judgment in determining whether GI contributes to or detracts from long-term urban resilience.

Integrating GI within broader resilience and adaptation strategies is central to this planning judgment. In arid cities, GI cannot compensate for urban forms that intensify heat exposure, disperse development, or increase infrastructure footprints. Instead, it must operate alongside planning strategies that address density, land-use efficiency, mobility patterns, and passive climate responsiveness. When GI is embedded within such strategies, it can reinforce resilience by enhancing the performance of urban systems as a whole. When it is isolated from them, it risks functioning as a remedial measure that masks deeper structural vulnerabilities.

This integration requires a shift from policy-driven greening toward planning-led adaptation. National and municipal policy frameworks increasingly recognize GI as part of the resilience discourse and acknowledge the importance of adaptation in urban contexts. Such recognition is valuable in elevating GI within planning agendas and legitimizing adaptation-oriented approaches. However, policy recognition alone does not determine planning outcomes. The critical issue lies in how these policy aspirations are translated into planning instruments, spatial decisions, and institutional practices at the urban scale.

In the Saudi context, national climate policy explicitly acknowledges adaptation and positions urban areas as key sites for responding to climate stress. This policy framing confirms that GI is understood, at least at the strategic level, as part of a broader adaptation agenda. Yet the existence of this recognition should not be conflated with effective implementation. The gap between policy ambition and planning practice remains a defining challenge, particularly in arid environments where environmental constraints expose weaknesses in coordination, governance, and resource management.

This gap highlights the need to move beyond symbolic greening toward context-sensitive planning tools. Symbolic greening emphasizes visibility, scale, and alignment with sustainability narratives, often prioritizing form over function. In contrast, context-sensitive planning treats GI as a conditional and selective intervention whose appropriateness depends on environmental limits, institutional capacity, and planning objectives. In arid cities, this means accepting that less visually "green" environments may, in fact, be more resilient if they reduce water demand, limit heat exposure, and align with long-term resource availability.

For planners, this shift carries several important implications. First, planners must adopt a critical stance toward GI proposals, evaluating them not only in terms of environmental intent but also in relation to water use, maintenance requirements, and integration with urban form. Second, GI should be prioritized in locations where its resilience benefits are most pronounced, such as dense neighborhoods, key public spaces, and pedestrian corridors, rather than applied uniformly across the city. Third, planning decisions must consider lifecycle performance, ensuring that GI can be sustained institutionally and financially under arid conditions.

For policymakers, the implications are equally significant. While policy frameworks play an important role in setting adaptation priorities, they must be accompanied by mechanisms that support implementation within planning systems. This includes aligning adaptation goals with land-use regulation, infrastructure investment, and institutional responsibilities. In the absence of such alignment, GI risks remaining a policy aspiration rather than a resilient urban outcome. Policymakers must therefore recognize that resilience is not delivered through declarations or programs alone, but through the cumulative effects of planning decisions over time.

Urban institutions also play a critical role in shaping the resilience outcomes of GI. Effective GI implementation in arid cities depends on coordination between planning authorities, water agencies, municipal services, and environmental regulators. Where institutional fragmentation persists, GI initiatives may suffer from unclear responsibilities, inconsistent maintenance, and limited accountability. Strengthening institutional coordination and clarifying governance arrangements are therefore essential components of a planning-led approach to urban resilience.

Taken together, these implications point toward a reframing of GI within arid urban planning. Rather than being treated as a default component of sustainable urbanism, GI should be understood as a conditional resilience tool whose value depends on context, scale, and integration. In Saudi Arabia, where environmental constraints are severe and policy ambition is high, this reframing is particularly important. It enables planners and policymakers to move beyond aspirational greening toward more grounded and resilient planning outcomes.

In summary, GI can enhance urban resilience in arid cities when it is selective, context-sensitive, and embedded within broader planning and adaptation strategies. Conversely, it becomes an unsustainable burden when driven by symbolic or policy-led imperatives that overlook environmental limits and institutional capacity. The Saudi experience underscores the importance of planning-led adaptation that aligns policy recognition with implementation realities. By positioning GI within this critical framework, the chapter reinforces the need to rethink resilience not as a matter of visibility or ambition, but as a function of planning integration, environmental realism, and long-term urban performance.

6.7 Conclusion: Toward Context-Sensitive Green Infrastructure Planning

This chapter has examined GI through the lens of arid urban environments, using Saudi cities to interrogate its role in climate adaptation and urban resilience. Rather than treating GI as an inherently beneficial or universally applicable planning solution, the chapter has emphasized the need to critically reassess GI within contexts defined by extreme heat, water scarcity, and environmental constraints. In doing so, it has highlighted the limitations of transferring GI models developed in temperate regions to desert cities without careful adaptation to local conditions.

A central conclusion emerging from this analysis is the inadequacy of universal GI models when applied to arid urban environments. Approaches that assume abundant water availability, low maintenance costs, and favorable climatic conditions risk undermining resilience when deployed in desert contexts. In Saudi cities, where water scarcity and energy intensity are structural characteristics of urban systems, such models can transform GI from a resilience-enhancing strategy into a long-term planning liability. Recognizing these limits is a necessary first step toward more realistic and effective approaches to urban adaptation.

This recognition underscores the necessity of arid-context planning approaches. Planning in desert environments requires strategies that explicitly account for environmental limits rather than seeking to overcome them through resource-intensive interventions. In the case of GI, this means prioritizing selectivity, efficiency, and integration over scale and visibility. Context-sensitive planning acknowledges that resilience in arid cities may depend less on the extent of green cover and more on the careful alignment of urban form, shading, density, and infrastructure coordination. Such an approach challenges dominant narratives of urban sustainability while offering a more grounded pathway toward long-term resilience.

Throughout the chapter, GI has been positioned not as a stand-alone solution, but as one component within a wider urban resilience toolkit. In arid contexts, GI can contribute meaningfully to resilience when combined with complementary strategies such as compact urban form, passive cooling, shaded public spaces, and efficient land-use planning. Conversely, when GI is expected to compensate for underlying vulnerabilities created by dispersed development patterns or poorly integrated planning decisions, its effectiveness is sharply limited. This insight reinforces the importance of viewing resilience as an emergent property of urban systems rather than as the outcome of isolated interventions.

The Saudi experience illustrates both the potential and the constraints of GI within rapidly urbanizing arid environments. National climate and adaptation frameworks recognize the importance of resilience and adaptation and elevate the role of urban areas in responding to climate stress. However, as the chapter has shown, policy recognition does not automatically translate into resilient planning outcomes. The effectiveness of GI ultimately depends on how adaptation ambitions are translated into land-use regulation, infrastructure coordination, and institutional practice. This translation gap highlights the central role of planning in mediating between environmental ambition and environmental reality.

By framing Saudi cities as policy-ambitious but environmentally constrained urban laboratories, the chapter contributes to the broader discussion within this book on rethinking urbanization, planning, and resilience. Rather than presenting Saudi cities as models to be replicated, the chapter uses their extreme conditions to surface critical questions about the assumptions embedded in contemporary planning paradigms. These insights extend beyond the Saudi context, offering lessons for other arid and semi-arid regions confronting rapid urban growth under climate stress.

Importantly, the analysis presented in this chapter aligns with the book's broader argument that resilience must be understood as a planning challenge rather than a purely environmental or technological one. GI, when treated as a symbolic marker of sustainability or a policy-driven objective, risks obscuring the deeper planning decisions that shape urban vulnerability and adaptive capacity. When approached as a planning instrument, embedded within land-use systems, governance structures, and environmental constraints, GI can play a more measured and constructive role in enhancing urban resilience.

This chapter also sets the stage for the discussion that follows. While GI represents one pathway for addressing climate adaptation at the scale of urban form and public space, it operates within broader trajectories of urban transformation. The next chapter shifts attention to megaproject-led urban development, examining how large-scale projects reshape planning priorities, resource allocation, and resilience outcomes in Saudi Arabia. In contrast to the selective and context-sensitive approach advocated for GI, megaprojects often operate at a scale that amplifies environmental and institutional risks. Understanding how these different modes of urban intervention interact is essential for developing a coherent planning framework capable of supporting resilient urban futures in arid environments.

In concluding, this chapter argues for a recalibration of GI planning in arid cities, one that moves beyond universal models and symbolic greening toward context-sensitive, planning-led adaptation. Such a recalibration is not only necessary for Saudi Arabia but for desert cities globally as they navigate the complex intersection of urbanization, climate stress, and resource constraint.

References

Almulhim, A. I., & Abubakar, I. R. (2023). Developing a sustainable water conservation strategy for Saudi Arabian cities. *Groundwater for Sustainable Development, 23*, 101040. https://doi.org/10.1016/j.gsd.2023.101040

Almulhim, A. I., & Cobbinah, P. B. (2023). Urbanization-environment conundrum: An invitation to sustainable development in Saudi Arabian cities. *International Journal of Sustainable Development & World Ecology, 30*(4), 359–373. https://doi.org/10.1080/13504509.2022.2152199

Almulhim, A. I., & Cobbinah, P. B. (2024a). Framing resilience in Saudi Arabian cities: On climate change and urban policy. *Sustainable Cities and Society, 101*, 105172. https://doi.org/10.1016/j.scs.2024.105172

Almulhim, A. I., & Cobbinah, P. B. (2024b). Urbanization provocateur: Reaching urban planning-led development in Saudi Arabia. *Land Use Policy, 147*, 107365. https://doi.org/10.1016/j.landusepol.2024.107365

Almulhim, A. I., & Cobbinah, P. B. (2025). Urban resilience and climate adaptation in countries of the Gulf cooperation council. In *Edward Elgar publishing eBooks* (pp. 122–139). https://doi.org/10.4337/9781803926926.00013

Amores, T. R. P., Ramos, J. S., Delgado, M. G., Medina, D. C., Cerezo-Narvaéz, A., & Domínguez, S. Á. (2023). Effect of green infrastructures supported by adaptative solar shading systems on livability in open spaces. *Urban Forestry & Urban Greening, 82*, 127886. https://doi.org/10.1016/j.ufug.2023.127886

Benedict, Mark & MacMahon, Edward. (2002). Green infrastructure: Smart conservation for the 21st century.

Dogan, R. (2026 February 18). King Salman Park advances toward 2026 opening on former Riyadh Airbase. ArchDaily. Retrieved February 19, 2026, from https://www.archdaily.com/1038859/king-salman-park-advances-toward-2026-opening-on-former-riyadh-airbase. ISSN 0719–8884.

Gao, K., Haddad, S., Paolini, R., Feng, J., Altheeb, M., Mogirah, A. A., Moammar, A. B., & Santamouris, M. (2024). The use of green infrastructure and irrigation in the mitigation of urban heat in a desert city. *Building Simulation, 17*(5), 679–694. https://doi.org/10.1007/s12273-024-1110-0

Government of Saudi Arabia. (2025). Second Nationally Determined Contribution (NDC) Report.

Lafortezza, R., Davies, C., Sanesi, G., & Konijnendijk, C. (2013). Green infrastructure as a tool to support spatial planning in European urban regions. *iForest—Biogeosciences and Forestry, 6*(3), 102–108. https://doi.org/10.3832/ifor0723-006

Liu, L., Palaiologou, F., & Schmidt-Iii, R. (2025b). Green infrastructure morphology: Integrating urban morphological perspectives for sustainable urban development. *Journal of Urban Management, 14*(4), 1231–1252. https://doi.org/10.1016/j.jum.2025.05.004

Liu, T., Wang, Y., Zhang, L., Xu, N., & Tang, F. (2025a). Outdoor thermal comfort research and its implications for landscape architecture: A systematic review. *Sustainability, 17*(5), 2330. https://doi.org/10.3390/su17052330

Meerow, S., Pajouhesh, P., & Miller, T. R. (2019). Social equity in urban resilience planning. *Local Environment, 24*(9), 793–808. https://doi.org/10.1080/13549839.2019.1645103

Mell, I., & Clement, S. (2019). Progressing green infrastructure planning: Understanding its scalar, temporal, geo-spatial and disciplinary evolution. *Impact Assessment and Project Appraisal, 38*(6), 449–463. https://doi.org/10.1080/14615517.2019.1617517

Monteiro, R., Ferreira, J., & Antunes, P. (2020). Green infrastructure planning principles: An integrated literature review. *Land, 9*(12), 525. https://doi.org/10.3390/land9120525

Moscatelli, M., & Raffa, A. (2023). Green infrastructure in arid urban contexts. Transitioning ecologies beyond green Riyadh. *AGATHÓN—International Journal of Architecture, Art and Design, 13*, 75–86. https://doi.org/10.19229/2464-9309/1362023

Okour, Y., & Shaweesh, H. (2024). Identifying the barriers to green infrastructure implementation in semi-arid urban areas using the DPSIR framework: A case study of Amman, Jordan. *City and Environment Interactions, 24*, 100165. https://doi.org/10.1016/j.cacint.2024.100165

Radonic, L., Galindo, V., Hanshaw, K., & Sandoval, F. (2025). Accessible, and culturally responsive: Why we need to examine diverse plant uses and values in green infrastructure. *Landscape and Urban Planning, 257*, 105317. https://doi.org/10.1016/j.landurbplan.2025.105317

Ruan, T., Xu, Y., Jones, L., Boeing, W. J., & Calfapietra, C. (2023). Green infrastructure sustains the food-energy-water-habitat nexus. *Sustainable Cities and Society, 98*, 104845. https://doi.org/10.1016/j.scs.2023.104845

Siddiqui, A., Maske, A. B., Khan, A., Kar, A., Bhatt, M., Bharadwaj, V., Kant, Y., & Hamdi, R. (2025). An urban climate paradox of anthropogenic heat flux and urban cool Island in a semi-arid urban environment. *Atmosphere, 16*(2), 151. https://doi.org/10.3390/atmos16020151

Part III
Governance, Transformation, and Urban Futures

Chapter 7
Managing Megaprojects and Building Resilience in Saudi Arabian Cities

7.1 Introduction

Over the past two decades, megaprojects have re-emerged as a defining feature of contemporary urban development, operating at increasingly large spatial and financial scales (Flyvbjerg, 2014). Defined as large-scale, complex, and transformational ventures that reshape cities and affect millions of people, megaprojects constitute not merely expanded versions of conventional projects, but a distinct category of urban intervention characterized by scale, ambition, and institutional complexity (Flyvbjerg, 2017; Kennedy, 2014). Empirical research consistently highlights systemic performance challenges associated with such projects, including cost overruns, schedule delays, and benefit shortfalls, patterns often described as the "iron law of megaprojects" (Flyvbjerg, 2017). Together, these characteristics situate megaprojects at the intersection of governance, performance risk, and long-term urban sustainability.

The contemporary resurgence of megaprojects is closely tied to intensified inter-urban competition and growth-oriented development strategies. Large-scale urban projects have increasingly become instruments of city repositioning within global economic hierarchies. This revival has been widely associated with neoliberal urban restructuring, in which states play strategic roles in facilitating capital investment through public–private partnerships and entrepreneurial planning frameworks (Orueta & Fainstein, 2008). Within this model, megaprojects function not only as spatial interventions but as vehicles of urban policy, embedding growth-oriented objectives within new governance arrangements.

Megaprojects are often legitimized through discourses of sustainability, social inclusion, and urban regeneration. Environmental improvements, public amenities, and social housing components are frequently incorporated into project narratives to secure political and public support. However, such elements may function as forms of "sweetening," masking growth-driven objectives behind socially and environmentally progressive rhetoric (Orueta & Fainstein, 2008). This tension between

A. I. Almulhim and P. B. Cobbinah, *Urbanization Rethought*, Urban Sustainability,
https://doi.org/10.1007/978-981-92-0118-1_7

stated public benefits and underlying development logics underscores the political dimensions of megaproject-led urban transformation.

From a planning perspective, the rise of megaprojects signals a significant shift in how urban change is conceptualized and implemented. Traditional planning paradigms are typically grounded in gradual development, regulatory continuity, and incremental spatial coordination. Megaprojects, by contrast, compress timeframes, centralize decision-making, and introduce step-change transformations that can rapidly alter urban form, infrastructure systems, and institutional arrangements (Flyvbjerg, 2014). In rapidly transforming contexts such as Saudi Arabia, this shift toward megaproject-led urbanism has been particularly pronounced, reflecting broader changes in development models and state-led planning practices (Alsayel et al., 2025).

Saudi Arabia provides a revealing lens through which to examine the implications of megaprojects for urban resilience. Urban development in the Saudi Arabia has entered a phase characterized by large-scale, state-led transformation initiatives that extend beyond isolated flagship projects to encompass comprehensive spatial restructuring agendas. Recent analyses describe this model as a form of accelerated, top-down urban transformation driven by strategic national repositioning objectives (Alsayel et al., 2025). These interventions frequently operate at metropolitan or regional scales, integrating land use, infrastructure systems, and governance mechanisms within unified development frameworks. As a result, megaprojects in Saudi Arabia do not merely introduce new districts but actively reshape planning priorities and institutional coordination structures.

The relevance of megaprojects to urban resilience lies not only in their physical scale, but also in their capacity to shape how cities respond to environmental, social, and institutional stress. Research on megaproject resilience emphasizes that resilience does not emerge automatically from scale or technological sophistication; rather, it depends on adaptive governance, stakeholder coordination, and institutional learning within complex project environments (Morkan et al., 2023). While large-scale interventions may create opportunities for integrated infrastructure provision and coordinated land-use planning, their size and resource intensity can also amplify vulnerability where governance mechanisms are insufficient. In environmentally constrained settings such as Saudi Arabia, where extreme heat, rapid urbanization, and ecological pressures increasingly shape development trajectories, the resilience implications of megaproject-led transformation become particularly acute (Almulhim & Cobbinah, 2025).

In arid environments, the stakes associated with large-scale urban transformation are heightened. Rapid land-use change and ecosystem disruption can intensify surface temperatures, strain infrastructure systems, and exacerbate environmental stress (Almulhim & Cobbinah, 2025). Megaprojects operating at scale can either reinforce these pressures or provide structured opportunities to address them more systematically. The determining factor lies not in the existence of megaprojects themselves, but in how they are planned, governed, and integrated within broader urban systems.

The Saudi planning context further complicates this dynamic through its institutional characteristics. Centralized decision-making structures and accelerated implementation timelines allow megaprojects to be conceived and executed at speeds rarely observed elsewhere. Recent analyses of urban transformation in Riyadh describe this model as a form of state-led "fast urbanism," characterized by top-down strategic direction, rapid project rollout, and large-scale spatial reconfiguration aimed at repositioning the city globally (Alsayel et al., 2025). While such capacity enables swift urban change, it also places significant pressure on planning institutions to ensure social inclusion, environmental sensitivity, and long-term adaptability. In this context, urban resilience cannot be assumed as an automatic by-product of scale or ambition, but must be actively embedded within planning processes capable of managing speed, centralization, and structural transformation.

This chapter approaches megaprojects as a planning phenomenon rather than as a collection of individual developments. It deliberately avoids descriptive accounts of specific projects in order to focus on the structural implications of megaproject-led urbanism for planning systems and resilience outcomes. By doing so, the chapter aligns with the book's broader objective of rethinking urbanization in Saudi Arabia through the lens of planning and resilience, rather than through policy narratives or project branding.

Central to this analysis is the recognition that megaprojects operate at a new urban scale that challenges existing planning frameworks. Their spatial footprints often extend beyond municipal boundaries, their infrastructure demands exceed conventional service provision models, and their governance arrangements may bypass established planning institutions. In the Saudi context, these dynamics are amplified by exceptional investment scales and state-led coordination of urban transformation (Alsayel et al., 2025). Together, these characteristics raise fundamental questions about accountability, coordination, adaptability, and long-term urban performance.

At the same time, megaprojects should not be viewed as inherently incompatible with urban resilience. Under certain conditions, their scale can enable integrated planning approaches that are difficult to achieve through incremental development. Comprehensive infrastructure provision, coordinated land-use planning, and the incorporation of climate-responsive systems at the outset may offer advantages in arid environments where retrofitting resilience is costly and complex. The challenge lies in ensuring that these potential advantages translate into durable planning outcomes rather than short-lived achievements.

This tension between opportunity and risk frames the core inquiry of the chapter. Rather than asking whether megaprojects are good or bad for cities, the chapter asks under what conditions megaprojects can support resilient urban outcomes in arid contexts such as Saudi Arabia, and when they risk becoming sources of long-term vulnerability. This conditional framing reflects the broader argument of the book: that resilience is not inherent in urban forms or development models, but emerges from the interaction between planning decisions, environmental constraints, and institutional capacity.

The structure of the chapter reflects this analytical focus. Following this introduction, Section 7.2 situates megaprojects within planning theory and urban resilience discourse, outlining key debates and conceptual tensions. Section 7.3 positions Saudi Arabia as a megaproject-driven urban context, highlighting the institutional and spatial features that shape planning outcomes. Section 7.4 examines the core tensions and trade-offs between megaproject development and urban resilience. Section 7.5 explores the implications of megaprojects under arid and climate-constrained conditions, building directly on the arguments developed in Chap. 6. Section 7.6 addresses governance and planning capacity challenges, while Section 7.7 synthesizes these insights to rethink megaprojects as planning instruments for resilience.

By framing megaprojects as both powerful catalysts of urban transformation and critical tests of planning systems, this chapter advances a more nuanced understanding of their role in shaping resilient cities. In the Saudi context, where environmental constraints intersect with large-scale development ambition at unprecedented scales, this analysis offers insights that are not only nationally relevant but also applicable to other regions confronting megaproject-led urbanization under conditions of climate stress.

7.2 Megaprojects in Planning Theory and Urban Resilience Discourse

Megaprojects have long occupied an ambivalent position within planning theory. On the one hand, they are frequently portrayed as engines of economic growth, urban competitiveness, and strategic urban transformation within market-oriented development frameworks (Durrant, 2025; Orueta & Fainstein, 2008). On the other, they are widely criticized for their tendency to bypass established planning processes, concentrate decision-making power, and generate persistent social, environmental, and financial risks (Flyvbjerg, 2017; Kennedy, 2014). This tension has become more pronounced as megaprojects increasingly intersect with contemporary debates on urban resilience, particularly in contexts characterized by environmental constraint and rapid urban change, where questions of governance capacity, institutional coordination, and adaptive management become central to long-term project performance (Lewin et al., 2023; Morkan et al., 2023).

Within the broader literature on megaproject management and planning, megaprojects are commonly understood as interventions that exceed conventional thresholds of scale, cost, and complexity (Flyvbjerg, 2017). Their defining feature is not merely size, but their capacity to restructure urban systems by introducing new spatial logics, infrastructure networks, and governance arrangements (Kennedy, 2014). Unlike incremental development, which allows planning systems to adjust gradually, megaprojects compress spatial and temporal scales, often requiring planning decisions to be made under conditions of uncertainty (Flyvbjerg, 2014). This compression challenges core planning principles related to participation, adaptability, and long-term coordination (Kennedy, 2014).

The relationship between megaprojects and urban resilience is therefore inherently complex. Resilience theory emphasizes adaptive capacity, system interdependence, and the ability of infrastructure systems to anticipate, absorb, and recover from shocks over time (Lewin et al., 2023). Megaprojects, by contrast, are often associated with structural rigidity, path dependency, and long-term infrastructural lock-in effects embedded within growth-oriented development trajectories (Durrant, 2025). Once implemented, their physical and institutional configurations can be difficult to modify, constraining future adaptation and reinforcing systemic risk (Flyvbjerg, 2017). This apparent contradiction has led scholars to question whether megaprojects can meaningfully contribute to resilience, or whether represent a mode of urban intervention marked by governance tensions and institutional complexity (Kennedy, 2014).

At the same time, recent research on infrastructure and project resilience indicates that resilience outcomes depend on governance arrangements, stakeholder coordination, and adaptive mechanisms that enable systems to respond to evolving risks (Lewin et al., 2023; Morkan et al., 2023). From this perspective, megaproject resilience depends not only on project design characteristics, but on stakeholder coordination processes and adaptive mechanisms that support collective responses to unexpected events (Morkan et al., 2023). While megaprojects may generate structural rigidity and long-term infrastructural commitments (Durrant, 2025), their future trajectories remain contingent on institutional arrangements and governance dynamics. This debate reflects a shift from viewing megaprojects as inherently problematic to assessing them as conditional planning instruments whose outcomes depend on governance design, stakeholder engagement, and implementation capacity.

These theoretical debates take on particular significance in arid and climate-constrained environments. In such contexts, resilience is not an abstract concept, but a material condition shaped by extreme heat, water scarcity, and ecological constraints (Almulhim & Cobbinah, 2025). Megaprojects operating at scale can either exacerbate these constraints or provide platforms for more integrated responses. Planning theory alone cannot resolve this tension; it must be interpreted through the lens of specific environmental and institutional settings. This is where the Saudi context becomes analytically central rather than merely illustrative.

Saudi Arabia exemplifies a planning environment in which megaprojects are not peripheral but central to urban transformation strategies (Alsayel et al., 2025). Unlike contexts where megaprojects emerge through complex negotiations among public and private actors, Saudi megaprojects are predominantly state-led and embedded within centralized governance structures that enable accelerated implementation and strategic spatial restructuring (Alsayel et al., 2025). This institutional configuration contrasts with assumptions commonly found in Western megaproject literature, which often emphasize fragmented authority, negotiated governance, and market-driven urban restructuring (Kennedy, 2014). In the Saudi context, questions of coordination, implementation capacity, and long-term institutional alignment become more analytically salient than concerns over speculative governance fragmentation, reshaping how megaprojects are interpreted within planning theory.

Urban resilience discourse emphasizes adaptation, transformation, and the capacity of urban systems to respond constructively to disturbance and uncertainty (Almulhim & Cobbinah, 2025). In the Saudi context, however, resilience governance operates within a predominantly centralized and top-down planning structure, where strategic decisions are concentrated at higher administrative levels (Almulhim & Cobbinah, 2024). While such centralization can facilitate coordination across sectors and scales, supporting the integration of infrastructure, land use, and environmental systems, it may also limit opportunities for distributed feedback, institutional learning, and iterative adjustment, which are central to resilience thinking.

Megaprojects further complicate this dynamic. Typically characterized as large-scale, long-term ventures requiring substantial financial and organizational commitments (Flyvbjerg, 2017), they embed socio-material resources in ways that generate path dependencies and structural lock-in (Durrant, 2025). These long-term infrastructural commitments can constrain flexibility over time, sitting uneasily alongside resilience principles that prioritize redundancy, incremental adjustment, and adaptive governance under conditions of uncertainty. In arid environments such as Saudi Arabia, where extreme heat, water scarcity, and ecological limits constitute structural rather than episodic conditions, this tension becomes particularly pronounced (Almulhim & Cobbinah, 2024). Planning megaprojects for resilience therefore requires navigating the interaction between centralized coordination, long-term infrastructural commitments, and the need for continuous institutional adaptation in the face of cumulative environmental stress.

Beyond questions of governance and institutional flexibility, resilience discourse also draws attention to the scale at which adaptation is organized. Scholars caution that overly large and centralized interventions may heighten systemic vulnerability, as disruptions at scale can produce cascading failures across interconnected systems (Lewin et al., 2023). By concentrating resources, infrastructure, and strategic functions within single large-scale developments, megaprojects inherently amplify the potential consequences of disruption. In Saudi Arabia, smart city development is largely driven by large-scale, nationally coordinated initiatives, with expert assessments highlighting the predominance of centralized planning structures in decision-making processes (Almulhim & Aina, 2025).

At the same time, scale can also be a source of resilience if it enables system-wide coordination. In planning theory, this raises questions about whether megaprojects should be understood as singular entities or as platforms within broader urban systems. In Saudi Arabia, where megaprojects frequently intersect with existing cities and infrastructure networks, their resilience implications extend beyond project boundaries. Planning theory must therefore move beyond project-centric evaluations toward system-oriented assessments that consider how megaprojects interact with surrounding urban environments.

The Saudi case also challenges normative assumptions about participation and governance embedded in much of the megaproject and resilience literature. While participatory planning is widely regarded as a cornerstone of resilient governance, its forms and institutional expressions vary across contexts. In Saudi Arabia, expert assessments highlight the predominance of centralized, top-down planning structures and limited citizen influence in smart city governance (Almulhim & Aina, 2025). Within this governance model, resilience outcomes may depend less on formal participation mechanisms and more on institutional coordination capacity, technical expertise, and long-term stewardship. This does not negate the importance of governance, but rather reframes it in ways that are attentive to local institutional configurations and decision-making structures.

More importantly, positioning Saudi Arabia within megaproject and resilience discourse does not imply exceptionalism. Rather, it highlights how extreme conditions can clarify underlying dynamics that are present, though less visible, elsewhere. The Saudi experience foregrounds questions of scale, speed, and environmental constraint that challenge generalized theories of planning and resilience. By examining these dynamics, the chapter contributes to a more nuanced understanding of how megaprojects function as planning instruments under conditions of climate stress.

In synthesizing planning theory and resilience discourse, this section underscores a central argument of the chapter: megaprojects are neither inherently resilient nor inherently vulnerable. Their resilience outcomes are shaped by how planning systems manage scale, integrate environmental constraints, and maintain adaptive capacity over time. In the Saudi context, where megaprojects play a pivotal role in urban transformation, these questions are not theoretical but practical, with long-term implications for urban sustainability and resilience.

The next section builds on this conceptual foundation by examining Saudi Arabia as a megaproject-driven urban context. It moves from theory to context-setting, exploring how institutional structures, planning practices, and environmental constraints shape the conditions under which megaprojects are conceived and implemented.

7.3 Saudi Arabia as a Megaproject-Driven Urban Context

Riyadh represents a distinctive urban planning context in which large-scale, government-led initiatives have become central instruments of spatial transformation (Alsayel et al., 2025). Rather than emerging as exceptional interventions within an otherwise incremental planning system, these projects increasingly define the trajectory of urban restructuring in the capital. As documented by Alsayel et al. (2025), recent development strategies rely on ambitious, state-driven programs that reshape urban form, infrastructure provision, and institutional arrangements at unprecedented scale and speed. This acceleration reflects broader economic repositioning and strategic spatial ambitions associated with national development agendas. In such a context, megaprojects function not merely as supplementary development tools but as primary vehicles for implementing comprehensive urban change within compressed implementation horizons. From a planning perspective, this shift alters the conditions under which decisions are made, emphasizing coordinated execution and strategic alignment at metropolitan scale. While this model enables rapid and integrated transformation, it may also limit opportunities for incremental adjustment and iterative learning when compared to more gradual planning systems (Alsayel et al., 2025).

The institutional configuration of urban planning in Saudi Arabia reinforces this megaproject-driven context. Urban development has historically been administered through a highly centralized framework, with planning authority concentrated at the national level and municipalities operating under strong central oversight (Aina et al., 2019). This structure positions the state as the primary actor in setting development priorities, mobilizing resources, and coordinating implementation. While recent reforms have introduced elements of decentralization, decision-making authority and financial control remain largely centralized. In such a system, large-scale urban initiatives are advanced through top-down governance mechanisms that align national vision, funding structures, and implementation processes (Aina et al., 2019).

This institutional environment has important implications for urban resilience. Centralized coordination can facilitate the integration of infrastructure systems, land-use planning, and environmental considerations within megaprojects (Aina et al., 2019). However, resilience frameworks emphasize the need for ongoing monitoring, predefined trigger points, and iterative adaptation over time (Lewin et al., 2023). When megaprojects are conceived as comprehensive and transformational interventions, their scale and embedded socio-material commitments may constrain flexibility and responsiveness to unforeseen environmental or socio-economic stresses (Durrant, 2025; Flyvbjerg, 2017).

Environmental context adds another layer of complexity to megaproject-led urbanization in Saudi Arabia. Cities operate under conditions of extreme heat, chronic water scarcity, and high energy demand, making environmental limits a structural feature of urban development rather than a peripheral concern (Almulhim & Cobbinah, 2024). Megaprojects implemented at scale interact directly with these constraints, amplifying both potential benefits and risks. Large-scale infrastructure can structure responses across interdependent systems and shape national resilience strategies (Lewin et al., 2023). However, megaprojects are also embedded within growth-oriented infrastructures whose material footprints and resource demands create long-term ecological consequences and path dependencies (Durrant, 2025). The Saudi context thus exposes the environmental stakes of megaproject planning more starkly than in less constrained settings.

Megaprojects often aim to reposition cities within global economic and cultural networks, contributing to processes of spatial restructuring that reshape existing urban patterns (Orueta & Fainstein, 2008). This transformation may generate differentiated spatial and governance dynamics, as large-scale, state-led developments reshape parts of the city through distinct planning and implementation frameworks (Alsayel et al., 2025). From a resilience perspective, such fragmentation raises concerns about the integration of interdependent systems and the coordination of infrastructure networks across urban regions, as resilience depends on managing linkages between interconnected systems and avoiding weaknesses along the chain of infrastructure provision (Lewin et al., 2023).

The relationship between megaprojects and existing cities is particularly important in the Saudi context. In many cases, megaprojects do not replace existing urban areas but intersect with them, influencing infrastructure flows, mobility patterns, and resource allocation. Planning systems must therefore manage not only the internal coherence of megaprojects but also their external interactions with surrounding urban environments. Failure to address these interactions can undermine resilience by creating mismatches between infrastructure capacity, service provision, and environmental performance across the urban system.

Planning capacity plays a critical role in mediating these dynamics. While Saudi Arabia possesses substantial technical and financial capacity to deliver large-scale urban initiatives (Aina et al., 2019; Alsayel et al., 2025), the institutional demands generated by multi-sectoral megaproject development are considerable. Coordinating land use, infrastructure provision, environmental management, and governance across project boundaries requires robust institutional integration and clear lines of responsibility (Aina et al., 2019). In practice, the accelerated pace and complexity associated with megaproject implementation may strain planning systems, increasing the risk of coordination gaps and long-term operational challenges.

These capacity challenges are compounded by the novelty and scale of megaproject-led urban transformation observed in Saudi Arabia (Alsayel et al., 2025). Planning institutions are required to adapt quickly to new modes of development that blur the boundaries between planning, project delivery, and governance. This adaptation process is ongoing and uneven, creating spaces of experimentation as well as vulnerability. From a resilience perspective, the ability of planning systems to evolve alongside megaprojects is as important as the design features of the projects themselves.

More significantly, framing Saudi Arabia as a megaproject-driven urban context does not imply uniformity across cities or developments. Rather, it highlights a common planning trajectory characterized by reliance on large-scale interventions to achieve rapid transformation. This trajectory reflects broader global trends, but it is intensified in the Saudi context by environmental constraints, centralized governance, and ambitious development agendas. These conditions make Saudi cities particularly valuable for examining how megaprojects interact with planning systems under stress.

Within the broader discourse on urban resilience, Saudi Arabia can be understood as an urban laboratory where the limits and possibilities of megaproject-led planning are tested under extreme conditions. The combination of high ambition, environmental constraint, and institutional capacity creates a context in which resilience outcomes are neither predetermined nor easily transferable. Instead, they emerge from the interaction between planning decisions, governance structures, and environmental realities over time.

This section has positioned Saudi Arabia as a megaproject-driven urban context that amplifies both the opportunities and risks associated with large-scale urban development. By situating megaprojects within this context, it provides the foundation for the critical analysis that follows. The next section builds on this groundwork by examining the specific tensions and trade-offs between megaproject development and urban resilience, focusing on how scale, resource intensity, and planning capacity shape resilience outcomes in Saudi cities.

7.4 Megaprojects and Urban Resilience: Tensions and Trade-Offs

Megaprojects occupy a paradoxical position in contemporary debates on urban resilience. Their scale, financial resources, and strategic visibility suggest significant potential to strengthen urban capacity and coordinate responses across interconnected infrastructure systems (Lewin et al., 2023). At the same time, megaprojects are characterized by high levels of uncertainty, risk exposure, and long-term socio-material commitments that can generate new vulnerabilities (Durrant, 2025; Flyvbjerg, 2017). In Saudi Arabia, where megaprojects have become central instruments of urban transformation (Alsayel et al., 2025), this paradox is especially pronounced. Understanding the relationship between megaprojects and urban resilience therefore requires a careful examination of the tensions and trade-offs embedded in large-scale planning decisions.

One of the most fundamental tensions concerns scale. Megaprojects operate at spatial and institutional scales that exceed those of conventional urban development. From a resilience perspective, scale introduces both transformative capacity and heightened exposure to risk. Large-scale projects mobilize substantial resources and reshape urban trajectories over extended time horizons. However, their scale also concentrates uncertainty and commits cities to long-term pathways that may prove difficult to adjust if underlying economic, environmental, or climatic assumptions change (Flyvbjerg, 2017).

In arid environments such as Saudi Arabia, urban development unfolds under structural environmental constraints characterized by extreme heat and chronic water scarcity (Almulhim & Cobbinah, 2024). When megaprojects are implemented at scale within such contexts, they interact directly with these pre-existing ecological limits. Large-scale transformation under conditions of resource stress heightens the importance of managing water, energy, and land systems in integrated ways. While megaprojects may incorporate advanced technologies and efficiency measures, their socio-material scale can entail substantial resource commitments and long-term infrastructural lock-in (Durrant, 2025).

Another central tension lies between speed and adaptability. Megaprojects are frequently justified on the basis of their ability to deliver rapid urban transformation. In the Saudi context, accelerated timelines are often viewed as a strategic advantage, enabling cities to respond quickly to economic and demographic change (Alsayel et al., 2025). However, resilience theory emphasizes the importance of learning, feedback, and incremental adjustment, processes that require time (Lewin et al., 2023). When development is compressed into short implementation cycles, opportunities for adaptation may be limited. This tension is particularly acute in environments where climate uncertainty is high and long-term environmental conditions are difficult to predict.

The issue of resource intensity further complicates the relationship between megaprojects and resilience. In desert cities, resilience is closely tied to the management of interdependent water and energy systems operating under conditions of scarcity and climatic stress (Almulhim & Cobbinah, 2025). The urban nexus perspective highlights how water provision, energy generation, and land-use planning are structurally interconnected, meaning that stress in one system can cascade across others. In Saudi Arabia, where water supply relies heavily on energy-intensive desalination processes, large-scale urban development increases dependence on tightly coupled infrastructure networks. While centralized systems may enhance operational reliability in the short term, they can also amplify systemic vulnerability by reducing redundancy and increasing exposure to resource and climate shocks. From a resilience standpoint, megaproject-led urbanization therefore raises critical questions about whether integrated infrastructure enhances adaptive capacity or deepens interdependence in ways that intensify long-term risk.

A further trade-off emerges in relation to innovation and lock-in. Megaprojects are often presented as ambitious, transformative interventions that introduce novel technologies and planning approaches (Flyvbjerg, 2017). However, their scale and complexity also entail long-term commitments that can be difficult to reverse once implementation is underway. Infrastructure networks, spatial configurations, and institutional arrangements established through megaprojects may become entrenched over time, creating path dependencies that constrain future adaptation if underlying environmental or economic assumptions change (Flyvbjerg, 2017).

Institutional dynamics play a critical role in shaping these trade-offs. In Saudi Arabia, urban development is characterized by strong centralization and dominant state involvement in planning and implementation processes (Aina et al., 2019). Such centralized authority can facilitate coordination across sectors and administrative scales, potentially supporting integrated planning approaches. However, it also concentrates decision-making power within national institutions and limits the influence of municipal actors and public participation. These governance configurations shape how planning priorities are defined and implemented, with implications for how resilience considerations are incorporated into large-scale development strategies.

Social dimensions of resilience intersect closely with megaproject-led development. Although megaprojects are often promoted as vehicles for economic growth and urban revitalization, their social impacts are uneven. Large-scale redevelopment initiatives have been criticized for producing regressive distributions of benefits and, in some cases, displacing lower-income populations (Orueta & Fainstein, 2008). Such outcomes raise concerns about social equity and inclusion, particularly where megaproject investment reshapes urban priorities and reallocates resources toward high-visibility development areas. In interdependent urban systems, weaknesses in one component can cascade across linked infrastructure networks, undermining overall system stability (Lewin et al., 2023).

Crucially, these tensions do not imply that megaprojects are inherently incompatible with urban resilience. Rather, they underscore the conditional nature of resilience outcomes. In Saudi Arabia, megaprojects can contribute to resilience when they are planned with explicit attention to environmental limits, resource efficiency, and long-term adaptability. Conversely, when megaprojects prioritize scale, speed, or symbolic impact without sufficient integration into planning systems, they risk generating new forms of vulnerability.

This conditionality places planning at the center of the megaproject–resilience relationship. Planning decisions shape how trade-offs between scale and flexibility, innovation and lock-in, and efficiency and redundancy are negotiated. In arid contexts, these decisions carry heightened significance due to the narrow margin for error imposed by environmental constraints. The Saudi experience illustrates how megaprojects can function as stress tests for planning systems, revealing whether institutions are capable of managing complexity and uncertainty at scale.

Ultimately, the tensions and trade-offs associated with megaproject-led urbanism challenge simplistic narratives of resilience. They demonstrate that resilience is not a by-product of ambition, investment, or technological sophistication. Instead, it emerges from the careful alignment of planning objectives, governance capacity, and environmental realities over time. In Saudi Arabia, where megaprojects play a central role in urban transformation, this alignment is both more difficult and more consequential than in less constrained contexts.

This section has highlighted the key tensions and trade-offs that shape the relationship between megaprojects and urban resilience in Saudi Arabia. By examining these dynamics, it provides a critical foundation for the subsequent sections, which explore how planning under arid and climate-constrained conditions and governance and implementation challenges further influence resilience outcomes.

7.5 Planning Megaprojects Under Arid and Climate-Constrained Conditions

Planning megaprojects in arid and climate-constrained environments entails fundamentally different risk structures compared to temperate contexts, as extreme heat exposure, chronic water scarcity, and high-climate-stress conditions operate as systemic pressures shaping infrastructure design, spatial planning, and long-term urban resilience (Hegazy et al., 2025). In desert regions, environmental constraints are not marginal variables to be mitigated through technical adjustments, but structural determinants that directly shape planning frameworks and adaptive capacity (Hegazy et al., 2025). In Saudi Arabia, where megaprojects have emerged as primary instruments of large-scale urban transformation, these environmental pressures intersect with geopolitical ambitions and post-oil economic repositioning, intensifying the strategic stakes of planning decisions and exposing the limitations of scale-driven development models (Kashef, 2025).

A defining feature of arid environments is chronic water scarcity, which fundamentally conditions urban development trajectories. In Saudi Arabia, extremely low precipitation rates, high evapotranspiration, declining groundwater reserves, and rapidly growing industrial and municipal demand have intensified reliance on desalination, which now supplies the majority of municipal water and remains highly energy-intensive (Almulhim & Abubakar, 2023). Megaprojects, by virtue of their size and infrastructural complexity, impose substantial demands on water systems during construction, operation, and long-term maintenance, thereby amplifying pressures on already constrained supply systems. In this context, urban development directly intersects with the water–energy nexus, where resource security depends on the interdependence between water production and energy consumption (Almulhim & Cobbinah, 2025). While integrated planning at the megaproject scale may create opportunities to optimize efficiency, the absolute scale of demand remains structurally high, raising concerns about long-term sustainability under intensifying climate stress.

Extreme heat constitutes a critical environmental constraint in Riyadh's urban context. Empirical analysis of land surface temperature between 2018 and 2022 indicates rising maximum temperatures and an expansion of urban heat island (UHI) areas, particularly in newly urbanized zones (Imam, 2023). The study further demonstrates that built-up areas and impervious surfaces are associated with elevated surface temperatures, contributing to increased thermal stress and higher energy demand for cooling. Notably, despite the implementation of the large-scale Green Riyadh urban greening initiative, UHI areas continued to expand during the study period, suggesting that the scale and intensity of vegetation interventions have not yet offset the thermal impacts of rapid urban expansion (Imam, 2023). In arid cities such as Riyadh, where extreme heat is persistent and intensifying, climate-sensitive design and sustained vegetation strategies remain central to mitigating thermal exposure and safeguarding urban liveability.

Energy demand constitutes a third critical dimension of climate-constrained planning, as urban systems are fundamentally structured around interconnected energy flows that sustain infrastructure, services, and economic activity (Rosales Carreón & Worrell, 2018). In arid cities, energy consumption is closely tied to cooling requirements, water production systems, and the continuous operation of urban infrastructure, reflecting the tight coupling between urbanization processes and rising energy demand (Johri et al., 2025). Megaprojects often concentrate these demands spatially and functionally, intensifying pressures on urban systems and interacting with the carrying capacity of cities (Zheng, 2020). While technological innovation may improve efficiency at the margin, the cumulative energy footprint of large-scale developments can remain substantial due to scaling effects within urban energy systems (Rosales Carreón & Worrell, 2018). From a resilience perspective, this concentration heightens dependence on continuous energy supply and increases vulnerability to systemic disruptions within interconnected infrastructure networks (Rosales Carreón & Worrell, 2018). Planning megaprojects under arid conditions therefore requires careful consideration of how energy demand scales with development size and how redundancy and flexibility can be embedded within urban infrastructure systems. The Riyadh Metro provides an illustration of how megaproject-scale infrastructure becomes embedded within tightly interconnected urban systems (see Fig. 7.1). While the metro enhances mobility efficiency and contributes to reducing automobile dependency, its operation remains structurally dependent on continuous energy supply, advanced cooling systems, and high-maintenance infrastructure within an extreme heat environment. In arid cities such as Riyadh, where summer temperatures regularly exceed thermal comfort thresholds, transport infrastructure cannot be understood independently of energy networks, cooling demand, and system-wide interdependencies. The resilience of such a megaproject therefore depends on both engineering robustness and the stability and adaptive capacity of the broader urban system within which it operates. In this sense, megaprojects do not eliminate climate exposure; they redistribute and concentrate it. These megaprojects resilience is therefore conditional upon governance capacity, infrastructure redundancy, and long-term operational adaptability.

The interaction between megaproject scale and climate risk is particularly significant. Large-scale developments amplify exposure to environmental hazards by concentrating assets, populations, and infrastructure within defined spatial footprints. In climate-constrained environments such as Saudi Arabia, rising temperatures, flood risks, and mounting infrastructure pressures illustrate how climate change intensifies urban vulnerability (Almulhim & Cobbinah, 2024). While megaprojects may incorporate advanced engineering solutions to manage such risks, their scale embeds infrastructures within tightly interconnected systems in which disruptions can propagate beyond their immediate site (Lewin et al., 2023). Moreover, large-scale infrastructure investments often create long-term path dependencies and structural lock-ins that limit adaptive flexibility when climate assumptions or resource conditions shift

Fig. 7.1 Riyadh Metro as a large-scale urban transport megaproject, *Source* (Saudipedia, 2025)

(Durrant, 2025). Under these conditions, failures are less likely to remain localized and may instead cascade across urban systems, raising critical questions about resilience planning in arid and climate-stressed regions (Lewin et al., 2023).

This raises a critical question for planning practice: can anticipatory planning at the megaproject scale compensate for environmental constraints, or does scale itself impose limits on adaptive capacity? Proponents of megaproject-led development argue that centralized and comprehensive planning enables the early integration of environmental considerations, infrastructure coordination, and regulatory oversight. However, the effectiveness of such integration depends on the extent to which environmental assessment mechanisms are institutionally embedded within planning frameworks and supported by appropriate legislative and administrative procedures (Aina et al., 2019). Moreover, large-scale infrastructure systems are characterized by interdependencies that can constrain adaptive flexibility and amplify disruption when climate risks exceed initial planning assumptions. In highly interconnected systems, failures are less likely to remain localized and may instead cascade across infrastructure networks, raising concerns about systemic vulnerability under conditions of environmental uncertainty (Lewin et al., 2023).

The relationship between scale, vulnerability, and resilience capacity is therefore central to understanding megaproject outcomes in arid environments. Larger scale can enable system-wide coordination and investment, potentially enhancing robustness. At the same time, highly integrated infrastructure systems can increase interdependence and amplify systemic vulnerability when disruptions occur (Lewin et al., 2023). Moreover, the material and financial magnitude of megaprojects may generate path dependencies that limit future flexibility and reduce opportunities for modular adaptation (Durrant, 2025). In Saudi Arabia, where megaprojects are frequently implemented through centralized and state-led frameworks that position them as integrated urban systems, the balance between coordination and flexibility becomes a critical planning concern (Alsayel et al., 2025). At the same time, escalating climate risks, including heat stress, flooding, and infrastructure strain, underscore the need for adaptive governance and evolving resilience strategies within Saudi urban contexts (Almulhim & Cobbinah, 2024). Planning frameworks must therefore address not only how megaprojects are designed, but how they can adjust and respond to shifting environmental conditions over time.

Importantly, arid and climate-constrained conditions challenge assumptions about the transferability of global megaproject models. Planning approaches developed in temperate or resource-abundant contexts often emphasize scale, iconic design, and intensive infrastructure provision. When applied in desert environments, these models may conflict with environmental realities and exacerbate vulnerability. The Saudi experience underscores the need to critically adapt megaproject planning paradigms to arid contexts rather than adopting them wholesale (Alsayel et al., 2025).

This critique does not suggest that megaprojects are inherently unsuitable for arid environments. Rather, it highlights the need for context-sensitive planning that recognizes environmental limits as foundational rather than peripheral. In Saudi Arabia, megaprojects that acknowledge water scarcity, thermal stress, and energy dependence as core planning parameters are more likely to contribute positively to resilience. Conversely, projects that treat these constraints as technical challenges to be overcome risk embedding long-term vulnerabilities within urban systems.

The analysis in this section reinforces the central argument developed in previous chapter: resilience in arid cities emerges from the alignment of planning decisions with environmental constraints, not from the scale or ambition of development alone. Megaprojects magnify both the potential and the risks of this alignment. Their scale can enable integrated responses to climate stress, but it can also amplify the consequences of misalignment. In climate-constrained environments such as Saudi Arabia, where margins for error are narrow, planning choices at the megaproject scale have far-reaching implications for urban resilience.

Overall, planning megaprojects under arid and climate-constrained conditions requires confronting fundamental questions about scale, vulnerability, and adaptive capacity. Water scarcity, extreme heat, and energy dependence shape the limits within which megaprojects operate, while their size intensifies exposure to climate risk. Anticipatory planning can mitigate some of these challenges, but it cannot eliminate environmental constraints or uncertainty. Understanding how megaproject scale interacts with climate vulnerability and resilience capacity is therefore essential for evaluating the role of megaprojects in shaping resilient urban futures in Saudi Arabia and other arid regions.

The following section builds on this analysis by examining governance, planning capacity, and implementation challenges, further unpacking how institutional arrangements influence the resilience outcomes of megaproject-led urban development.

7.6 Governance, Planning Capacity, and Implementation Challenges

Megaproject-led urban development places exceptional demands on governance systems and planning institutions, often operating through distinctive institutional arrangements and specialized regulatory regimes that extend beyond conventional planning mechanisms (Kennedy, 2014). While the physical scale and technical complexity of megaprojects are frequently emphasized, their resilience outcomes are equally shaped by governance structures, regulatory frameworks, and long-term operational capacity. In Saudi Arabia, where megaprojects have become central instruments of state-led urban transformation implemented through highly centralized governance frameworks, these institutional dimensions play a decisive role in determining whether large-scale developments translate ambitious planning visions into resilient urban realities (Alsayel et al., 2025).

A fundamental governance question in the context of megaprojects concerns who plans, who coordinates, and who implements. Unlike conventional urban development, which is typically guided by established municipal planning processes, megaprojects often involve hybrid governance arrangements that cut across administrative boundaries and institutional mandates (Orueta & Fainstein, 2008). Planning authority may be centralized within specialized entities, while coordination spans multiple ministries, agencies, and service providers. Implementation, in turn, frequently relies on complex networks of public and private actors. This fragmentation of roles complicates accountability and raises challenges for coherent, resilience-oriented planning.

In the Saudi context, centralized governance structures enable the mobilization of resources and alignment of strategic objectives at a national scale, facilitating the rapid initiation and delivery of large-scale urban projects (Aina et al., 2019). However, such centralization also reshapes the role of planning institutions, which increasingly shift from regulating incremental development to managing large-scale transformation processes. This transition places new demands on institutional capacity and long-term governance effectiveness, particularly in relation to sustainable and resilient urban development (Aina et al., 2019).

Institutional coordination emerges as one of the most significant challenges in this environment. Megaprojects require close alignment between land-use planning, infrastructure provision, environmental management, and service delivery (Flyvbjerg, 2017). In practice, these functions are often distributed across institutions with distinct priorities, timelines, and performance metrics, complicating coherent decision-making. When coordination mechanisms are weak or informal, resilience considerations can be sidelined in favor of immediate delivery objectives. In arid and climate-constrained contexts such as Saudi Arabia, the consequences of such misalignment are magnified, as failures in coordination can exacerbate climate vulnerability and infrastructure strain (Almulhim & Cobbinah, 2024).

Regulatory frameworks represent another critical dimension of governance capacity. Megaprojects frequently operate under customized regulatory regimes and distinctive governance arrangements that extend beyond conventional planning processes (Kennedy, 2014). While regulatory flexibility can enable experimentation and integrated planning, it can also create gaps in oversight and weaken the application of established planning standards. In the Saudi context, adapting regulatory frameworks to accommodate megaprojects raises questions about how resilience requirements are defined, enforced, and monitored over time. The challenge lies in balancing flexibility with accountability, ensuring that regulatory adaptation does not undermine long-term urban performance.

The issue of long-term operation and maintenance further highlights the institutional challenges associated with megaproject-led development. Planning processes often focus on vision, design, and delivery, while operational considerations receive less sustained attention. The resilience of complex and interdependent infrastructure systems depends heavily on long-term monitoring, maintenance, and institutional capacity to respond to emerging risks (Lewin et al., 2023). In arid environments such as Saudi Arabia, where urban systems operate under persistent climate stress, including heat, water scarcity, and infrastructure strain, the consequences of weak operational capacity are magnified (Almulhim & Cobbinah, 2024). Megaprojects introduce systems that require continuous institutional oversight to function effectively under extreme climatic conditions. Where responsibilities for operation and maintenance are unclear or under-resourced, resilience outcomes can deteriorate rapidly after initial delivery.

This gap between planning vision and operational resilience is a recurring challenge in megaproject governance, where large-scale projects often operate through distinctive institutional and political arrangements that prioritize strategic ambition and high-impact transformation (Kennedy, 2014). Planning visions tend to emphasize integrated design, technological innovation, and transformative impact. Operational resilience, by contrast, depends on routine practices, institutional coordination, and sustained resource allocation over time. In Saudi Arabia, where megaprojects are frequently associated with high-profile planning narratives, there is a risk that attention shifts away from the less visible but critical processes that underpin long-term resilience. Bridging this gap requires planning institutions to engage more directly with operational realities from the earliest stages of project conception.

Beyond the tension between vision and operation, institutional capacity constraints further shape resilience outcomes. Megaproject development frequently operates through complex governance arrangements that demand coordination, institutional alignment, and administrative capacity beyond conventional urban planning functions (Kennedy, 2014). Managing megaprojects requires skills and expertise that extend beyond traditional land-use planning, including systems integration, risk management, and inter-institutional coordination. While Saudi Arabia has invested significantly in technical and professional capacity, the scale and novelty of megaproject-led urbanism continue to test institutional limits. Rapid implementation timelines can further strain capacity, reducing opportunities for learning and adjustment. From a resilience perspective, institutional adaptability is as important as technical competence.

At the broader urban scale, urban governance challenges are also reflected in the relationship between megaprojects and existing city systems. Megaprojects present new governance arrangements that operate alongside established municipal structures. The Diriyah Gate Project (see Fig. 7.2), which is managed and implemented through the Diriyah Gate Development Authority, provides a contemporary case of this governance interface. Operating through a dedicated institutional framework, the project exemplifies how megaproject governance structures coexist alongside established municipal systems. While this arrangement facilitates strategic alignment and concentrated resource mobilization, it also necessitates careful coordination across planning, heritage management, infrastructure provision, and environmental regulation. In the absence of clearly defined institutional interfaces, governance fragmentation can undermine system-wide resilience, particularly in climate-constrained urban environments.

This coexistence can create ambiguity regarding authority, responsibility, and accountability, particularly in areas such as infrastructure integration and service provision. In Saudi cities, where megaprojects frequently intersect existing urban areas, managing these interfaces is critical for system-wide resilience. Failure to

Fig. 7.2 Diriyah gate project in Riyadh, illustrating the spatial integration of heritage conservation, mixed-use urban development, and large-scale governance coordination within an existing metropolitan context. *Source* (Diriyah Company, 2025)

coordinate governance across project and city boundaries can generate vulnerabilities that undermine both new and existing urban systems.

Importantly, these governance and capacity challenges do not imply that megaprojects are inherently unmanageable or incompatible with resilience objectives. Rather, they highlight the need for planning-led governance frameworks that prioritize long-term performance over short-term delivery. In arid contexts, where environmental constraints impose narrow margins for error, governance arrangements must support ongoing adaptation, monitoring, and institutional learning. This requires a shift from viewing megaprojects as finite delivery tasks toward understanding them as evolving urban systems.

The Saudi experience underscores the importance of aligning governance structures with the realities of climate-constrained urbanization. Centralized authority can support strategic coordination, but it must be complemented by mechanisms that enable cross-sector collaboration, regulatory coherence, and operational continuity. Planning institutions play a critical role in mediating between strategic ambition and practical implementation, translating planning visions into resilient outcomes over time.

In this sense, governance capacity becomes a central determinant of whether megaprojects enhance or undermine urban resilience. Strong planning institutions can harness the scale and resources of megaprojects to support integrated adaptation strategies. Weak coordination, insufficient regulatory oversight, or underdeveloped operational capacity can transform the same projects into sources of long-term vulnerability. Recognizing this conditionality is essential for moving beyond simplistic evaluations of megaproject-led urbanism.

This section has emphasized that resilience in the context of megaprojects is not delivered by design excellence or technological sophistication alone. It is produced through governance arrangements that align planning vision with implementation capacity and long-term operation. In Saudi Arabia, where megaprojects are reshaping urban futures at unprecedented scales, strengthening institutional coordination, regulatory adaptation, and operational resilience is critical for ensuring that ambitious planning visions translate into sustainable and resilient urban outcomes.

The next section builds on this institutional analysis by rethinking megaprojects as planning instruments for resilience, synthesizing the chapter's arguments and identifying the conditions under which megaproject-led development can contribute positively to resilient urban futures.

7.7 Rethinking Megaprojects as Planning Instruments for Resilience

The analysis developed in the preceding sections demonstrates that megaprojects cannot be understood as inherently resilient or inherently detrimental to urban systems. Rather, their resilience outcomes are shaped by how they are conceived, planned, governed, and integrated within broader urban and environmental contexts. This section rethinks megaprojects not as exceptional developments or symbolic interventions, but as planning instruments whose effectiveness depends on the conditions under which they operate. In arid and climate-constrained environments such as Saudi Arabia, this reframing is particularly critical.

Viewing megaprojects as planning instruments shifts attention away from their iconic form or economic promise and toward their functional role within urban systems. As planning instruments, megaprojects influence land-use patterns, infrastructure provision, governance arrangements, and resource flows at scale. They structure urban trajectories over long time horizons and shape the capacity of cities to adapt to environmental and socio-economic change. From a resilience perspective, this means that megaprojects should be evaluated not in isolation, but in terms of how they interact with existing cities, institutions, and environmental constraints.

A key implication of this reframing is the recognition that scale is not a substitute for resilience. While the scale of megaprojects can enable integrated planning and coordinated investment, it also magnifies the consequences of planning failure, as increasing size and complexity are closely associated with heightened systemic risk and performance uncertainty (Flyvbjerg, 2017). In Saudi Arabia, where megaprojects often operate at metropolitan or regional scales, planning decisions embed assumptions about water availability, energy supply, and climate conditions that may persist for decades. Rethinking megaprojects as planning instruments therefore requires explicit acknowledgment of uncertainty and the limits of predictive planning in arid environments.

One condition under which megaprojects can support resilience is when they are planned as phased and adaptive systems rather than as fixed, monolithic developments. Phasing allows planning institutions to incorporate learning, adjust to changing environmental conditions, and recalibrate development trajectories over time. In climate-constrained contexts, where long-term uncertainty is unavoidable, the capacity to adapt incrementally is essential. Megaprojects that lock cities into rigid spatial and infrastructural configurations risk undermining resilience by reducing flexibility and increasing exposure to unforeseen stress (Lewin et al., 2023).

Integration with existing urban systems represents another critical condition for resilience-oriented megaproject planning. When megaprojects are conceived as self-contained entities operating through distinctive governance and development regimes, they can reconfigure spatial priorities and institutional arrangements in ways that reshape metropolitan coherence (Orueta & Fainstein, 2008). In contrast, megaprojects that are planned as extensions or strategic reconfigurations of existing urban fabrics are more likely to reinforce connectivity and functional integration across metropolitan systems. In the Saudi context, where megaprojects often coexist with established cities, this distinction has significant implications for infrastructure performance, service provision, and environmental management.

Rethinking megaprojects also requires a shift from project-centric evaluation toward system-oriented planning. Traditional project evaluation frameworks tend to focus on delivery metrics, economic outputs, and short-term performance. Resilience-oriented planning demands a broader perspective that considers long-term operational capacity, environmental sustainability, and institutional adaptability. In arid environments, where resource constraints are persistent rather than episodic, such system-oriented evaluation is essential for avoiding development pathways that appear successful initially but become unsustainable over time.

Governance arrangements play a decisive role in determining whether megaprojects function as effective planning instruments for resilience. As discussed in the previous section, governance structures that prioritize delivery speed and centralized control may achieve rapid transformation but struggle to support long-term adaptation. Rethinking megaprojects for resilience involves aligning governance frameworks with the requirements of ongoing management, monitoring, and learning. This includes clarifying institutional responsibilities, strengthening coordination mechanisms, and embedding resilience considerations within regulatory and operational practices.

In the Saudi context, centralized governance offers both opportunities and constraints for resilience-oriented megaproject planning. Strong state capacity can facilitate coordination across sectors and scales, enabling integrated responses to environmental challenges. However, this capacity must be complemented by institutional mechanisms that support feedback, adjustment, and accountability over time. Without such mechanisms, megaprojects risk becoming rigid instruments that deliver initial outcomes but lack the flexibility required to respond to evolving climate and resource conditions.

Another important dimension of rethinking megaprojects concerns the relationship between planning ambition and environmental realism. Megaprojects are often driven by transformative visions that emphasize innovation, global competitiveness, and symbolic impact. While such ambition can mobilize resources and political support, it can also obscure environmental limits, particularly in arid regions.

Resilience-oriented planning requires tempering ambition with realism, ensuring that megaprojects are aligned with water availability, energy capacity, and long-term environmental constraints. This alignment is not a constraint on planning creativity, but a prerequisite for sustainable urban futures.

The Saudi experience underscores the importance of moving beyond policy-driven or vision-led megaprojects toward planning-led instruments grounded in environmental and institutional realities. Policy recognition of resilience and adaptation is a necessary starting point, but it does not guarantee resilient outcomes. The translation of policy ambition into planning practice depends on how megaprojects are embedded within land-use systems, infrastructure planning, and governance arrangements. Rethinking megaprojects as planning instruments foregrounds this translation process and highlights the role of planners in mediating between ambition and implementation.

This reframing also has implications for how success is defined and measured. In resilience-oriented planning, success cannot be assessed solely through completion milestones, economic indicators, or visual impact. Instead, it must be evaluated in terms of long-term performance, adaptability, and contribution to system-wide resilience. In arid contexts, where environmental stress is ongoing, the true test of megaprojects lies in their capacity to operate sustainably over time rather than in their initial delivery.

Rethinking megaprojects as planning instruments does not entail rejecting large-scale development altogether. Rather, it involves recognizing that scale amplifies both potential and risk. In Saudi Arabia, megaprojects will continue to play a significant role in shaping urban futures. The critical question is not whether megaprojects should exist, but how they can be planned, governed, and integrated in ways that support resilience rather than undermine it. This conditional perspective aligns with the broader argument of the book, which emphasizes planning as the key mediator between urbanization and resilience.

By situating megaprojects within a resilience-oriented planning framework, this section contributes to a more nuanced understanding of large-scale urban development in arid environments. It challenges deterministic narratives that portray megaprojects as either panaceas or pathologies, and instead highlights the contingent nature of their outcomes. In doing so, it reinforces the need for planning systems capable of managing complexity, uncertainty, and long-term environmental constraint.

Overall, rethinking megaprojects as planning instruments for resilience requires a shift in both analytical perspective and planning practice. It demands attention to scale, integration, governance, and adaptability, particularly in arid and climate-constrained contexts such as Saudi Arabia. When these conditions are met, megaprojects can contribute to resilient urban futures. When they are ignored, the same projects risk embedding long-term vulnerabilities within urban systems. This reframing sets the stage for the chapter's conclusion, which synthesizes the implications of megaproject-led urbanism for planning and resilience in Saudi Arabia and beyond.

7.8 Conclusion: From Iconic Development to Resilient Urban Futures

This chapter has examined megaproject-led urban development through the lens of planning and urban resilience, using Saudi Arabia as a critical context in which the implications of scale, speed, and ambition are particularly pronounced. Rather than approaching megaprojects as isolated symbols of transformation or as inherently beneficial urban interventions, the chapter has treated them as planning instruments whose outcomes are shaped by environmental constraints, institutional capacity, and governance choices. In doing so, it has highlighted the conditional nature of resilience in the context of large-scale urban transformation.

A central conclusion emerging from the chapter is that megaprojects do not inherently produce resilient urban futures. Their scale and resources can enable integrated planning and infrastructure provision, but they also amplify risk, intensify resource demands, and reduce adaptive flexibility when not carefully managed. In arid and climate-constrained environments such as Saudi Arabia, these dynamics are magnified. Environmental limits related to water scarcity, extreme heat, and energy dependence impose narrow margins for error, making the consequences of planning decisions at the megaproject scale both far-reaching and long-lasting.

The analysis has shown that the relationship between megaprojects and urban resilience is best understood in terms of tensions and trade-offs, rather than simple cause-and-effect relationships. Trade-offs between scale and adaptability, speed and learning, innovation and lock-in, and ambition and environmental realism recur throughout megaproject planning processes. These trade-offs are not unique to Saudi Arabia, but the Saudi context brings them into sharper focus due to the intensity of development ambition and the severity of environmental constraints. As such, Saudi cities function as revealing stress tests for megaproject-led urbanism under climate pressure.

One of the key insights of the chapter is the importance of planning-led approaches to megaproject development. Where megaprojects are driven primarily by policy vision, symbolic objectives, or delivery imperatives, resilience considerations risk being subordinated or treated as secondary design features. In contrast, when megaprojects are embedded within robust planning frameworks that prioritize environmental limits, system integration, and long-term adaptability, they can contribute more meaningfully to urban resilience. This distinction reinforces the broader argument of the book: that resilience is produced through planning processes rather than through scale, investment, or technology alone.

Governance and institutional capacity have emerged as decisive factors in shaping megaproject resilience outcomes. The chapter has demonstrated that ambitious planning visions must be supported by effective coordination, regulatory adaptation, and long-term operational capacity if they are to translate into resilient urban systems. In Saudi Arabia, centralized governance structures offer opportunities for strategic alignment and resource mobilization, but they also create risks related to rigidity, limited feedback, and institutional overload. Bridging the gap between planning vision and operational resilience requires sustained institutional learning, cross-sector coordination, and attention to everyday practices of urban management.

The chapter has also underscored the significance of environmental context in evaluating megaprojects. In arid environments, environmental constraints are not external challenges to be engineered away, but foundational conditions that must be incorporated into planning logic. Megaprojects that treat water scarcity, heat stress, and energy dependence as technical problems risk embedding long-term vulnerabilities within urban systems. By contrast, projects that recognize these constraints as planning parameters can better align development ambition with environmental reality. This insight builds directly on the arguments advanced in Chap. 6, reinforcing the need for context-sensitive approaches to resilience in desert cities.

Importantly, the chapter does not advocate for the rejection of megaprojects as a mode of urban development. Given their central role in Saudi Arabia's urban transformation, megaprojects will continue to shape urban futures for the foreseeable future. The critical issue is not whether megaprojects should be pursued, but how they are planned, governed, and integrated within wider urban systems. Rethinking megaprojects as planning instruments highlights the possibility of recalibrating their role, from iconic, stand-alone developments toward components of more resilient, adaptable urban frameworks.

This recalibration involves shifting the evaluation of megaprojects away from short-term delivery metrics and toward long-term performance. Resilient urban futures depend on how megaprojects function over time, how they respond to environmental stress, and how they interact with existing cities and institutions. In this sense, resilience is not an outcome achieved at project completion, but an ongoing process shaped by planning decisions, governance practices, and environmental change.

Within the broader structure of the book, this chapter complements earlier discussions on urbanization, planning, and resilience in Saudi Arabia by focusing on one of the most influential drivers of contemporary urban change. Together with Chap. 6, which examined green infrastructure under arid conditions, the chapter reinforces a central theme: that planning integration and environmental realism are essential for resilient urbanization in Saudi Arabia. While green infrastructure and megaprojects operate at different scales, both illustrate how resilience outcomes depend on context-sensitive planning rather than universal models or policy rhetoric.

The insights developed in this chapter also have relevance beyond the Saudi context. As cities worldwide increasingly turn to megaprojects to address economic restructuring, climate adaptation, and urban growth, the challenges identified here, particularly those related to scale, governance, and environmental constraint, are likely to become more widespread. Arid and semi-arid regions, in particular, can draw lessons from the Saudi experience about the risks of scale-driven development and the importance of aligning ambition with resource realities.

In concluding, this chapter argues for a shift from iconic development toward resilient urban futures. Such a shift does not diminish the transformative potential of megaprojects, but it reframes their purpose within a planning-led resilience agenda. In Saudi Arabia, where urban transformation is occurring at unprecedented scales and speeds, this reframing is both urgent and consequential. Megaprojects will shape the spatial, environmental, and institutional landscapes of Saudi cities for generations. Whether they contribute to resilience or embed long-term vulnerability will depend on the capacity of planning systems to manage complexity, uncertainty, and environmental constraint over time.

This chapter thus closes by reaffirming the central premise of the book: that rethinking urbanization in Saudi Arabia requires moving beyond singular solutions and toward integrated, context-sensitive planning approaches. Megaprojects, like green infrastructure, are neither solutions nor problems in themselves. They are instruments whose impacts are determined by planning choices. Understanding and reshaping those choices is essential for building resilient urban futures in Saudi Arabia and other regions confronting rapid urban transformation under climate stress.

References

Aina, Y. A., Wafer, A., Ahmed, F., & Alshuwaikhat, H. M. (2019). Top-down sustainable urban development? Urban governance transformation in Saudi Arabia. *Cities, 90*, 272–281. https://doi.org/10.1016/j.cities.2019.03.003

Almulhim, A. I., & Abubakar, I. R. (2023). Developing a sustainable water conservation strategy for Saudi Arabian cities. *Groundwater for Sustainable Development, 23*, 101040. https://doi.org/10.1016/j.gsd.2023.101040

Almulhim, A. I., & Aina, Y. A. (2025). Achieving human-centered smart city development in Saudi Arabia. *Urban Science, 9*(10), 393. https://doi.org/10.3390/urbansci9100393

Almulhim, A. I., & Cobbinah, P. B. (2025). Dammam metropolitan area: Advancing the pursuit of urban resilience. *Habitat International, 163*, 103499. https://doi.org/10.1016/j.habitatint.2025.103499

Alsayel, A., Fransen, J., De Jong, M., & Almatar, K. M. (2025). Riyadh: King of gigaprojects can Riyadh's dramatic place rebranding realize path-breaking and profound urban transformations? *Cities, 163*, 106022. https://doi.org/10.1016/j.cities.2025.106022

Carreón, J. R., & Worrell, E. (2018). Urban energy systems within the transition to sustainable development. A research agenda for urban metabolism. *Resources Conservation and Recycling, 132*, 258–266. https://doi.org/10.1016/j.resconrec.2017.08.004

Diriyah Company. (2025). *MasterPlan, Saudi Arabia.* https://www.diriyahcompany.sa/en/masterplan

Durrant, D. (2025). Megaprojects and the transition to post-growth infrastructure. *Economy and Society, 54*(4), 672–694. https://doi.org/10.1080/03085147.2025.2589653

Flyvbjerg, B. (2014). What you should know about megaprojects and why: An overview. *Project Management Journal, 45*(2), 6–19. https://doi.org/10.1002/pmj.21409

Flyvbjerg, B. (2017). Introduction. *Oxford University Press eBooks.* https://doi.org/10.1093/oxfordhb/9780198732242.013.1

Hegazy, I., Bahreldin, I., Maddah, R., Hammad, H., & Ibrahim, H. (2025). Adapting global urban resilience frameworks to the Saudi context: Toward climate-responsive policy solutions under vision 2030. *International Journal of Low-Carbon Technologies, 20*, 2089–2102. https://doi.org/10.1093/ijlct/ctaf139

Imam, A. (2023). Examining the impact of green Riyadh project on alleviating urban heat Island effects. *Journal of Umm Al-Qura University for Engineering and Architecture, 14*(4), 201–211. https://doi.org/10.1007/s43995-023-00030-4

Johri, S., Selvakumar, P., Satyanarayana, P., Das, A., Gupta, S., & Bhattacharya, S. (2025). Urbanization and energy: The nexus of global challenges. In F. Özsungur, M. Chaychi Semsari, &

H. Küçük Bayraktar (Eds.), *Geopolitical landscapes of renewable energy and urban growth* (pp. 453–478). IGI Global Scientific Publishing. https://doi.org/10.4018/979-8-3693-8814-3.ch016

Kashef, M. (2025). Rethinking Masdar and the line megaprojects: The interplay of economic, social, political, and spatial dimensions. *Land, 14*(7), 1358. https://doi.org/10.3390/land14071358

Kennedy, L. (2014). The politics and changing paradigm of megaproject development in metropolitan cities. *Habitat International, 45*, 163–168. https://doi.org/10.1016/j.habitatint.2014.07.001

Lewin, C., Rossi, M., Soultani, E., & Raj, K. S. (2023). Managing infrastructure resilience and adaptation. *Sustainable and Resilient Infrastructure, 9*(2), 107–123. https://doi.org/10.1080/23789689.2023.2241728

Morkan, B., Bertels, H. M., Sheth, A., & Holahan, P. J. (2023). Building megaproject resilience with stakeholders: The roles of citizenship behavior and critical transition mechanisms. *International Journal of Project Management, 41*(5), 102485. https://doi.org/10.1016/j.ijproman.2023.102485

Orueta, F. D., & Fainstein, S. S. (2008). The new mega-projects: Genesis and impacts. *International Journal of Urban and Regional Research, 32*(4), 759–767. https://doi.org/10.1111/j.1468-2427.2008.00829.x

Saudipedia. (2025). *Why is Riyadh Metro the backbone of the public transport system in the capital? Saudipedia.* https://saudipedia.com/en/why-is-riyadh-metro-the-backbone-of-the-public-transport-system-in-the-capital#

Zheng, L. (2020). Research on the impact of mega-projects on carrying capacity of cities taking the first-line project of the west-east gas pipeline as an example. *Journal of Management Science and Engineering, 5*(3), 195–211. https://doi.org/10.1016/j.jmse.2020.08.001

Chapter 8
Advancing the Pursuit of Urban Resilience: Profiling Dammam Metropolitan Area

8.1 Introduction

Building resilient cities has emerged as an important oasis of urban policy orthodoxy in managing climate change and addressing urban sustainability inadequacies. As the colliding impact of climate change and rapid urbanization unfolds at an unprecedented scale worldwide particularly in the Global South (Almulhim & Sharifi, 2025; Haddad et al., 2015; Hoekstra et al., 2018; Li et al., 2022; Raihan & Tuspekova, 2022; Song et al., 2018; Wigginton et al., 2016), urban resilience object of dealing with the implications of unplanned urbanization, mitigating and adapting to climate change, and promoting inclusive urban development has become a central policy ideal. Meerow et al., (2016, p. 39) explain urban resilience as

> The ability of an urban system-and all its constituent socio-ecological and socio-technical networks across temporal and spatial scales-to maintain or rapidly return to desired functions in the face of a disturbance, to adapt to change, and to quickly transform systems that limit current or future adaptive capacity.

These urban resilience ideals are reflected in major global efforts including the New Urban Agenda, UN Sustainable Development Goals, and the activities of the Global Commission on the Economy and the Climate, and the Intergovernmental Panel on Climate Change (IPCC). These global urban resilience efforts underscore the importance of the concept as a risk management and disaster reduction paradigm, which is central to mitigate recent severity of urban development pressures and climate change impacts, such as hurricanes, flood events, droughts, and urban sprawl (Li et al., 2021; Nagendra et al., 2018; Nhamo et al., 2023).

This chapter profiles the Dammam Metropolitan Area (DMA) to problematize dominant understandings of urban resilience and its relationship to urban sustainability. In academic research, the urban resilience idea has assumed a growing and significant importance in urban studies and environmental planning over the past decades (e.g., Breheny, 1992; Burton, 2000; Jenks et al., 1996; Næss, 2006). The core idea of urban resilience in planning gained significant momentum following

A. I. Almulhim and P. B. Cobbinah, *Urbanization Rethought*, Urban Sustainability,
https://doi.org/10.1007/978-981-92-0118-1_8

Mileti (1999) use of the concept for urban disaster risk management. Since then, urban resilience has evolved to reflect an urban development initiative that responds to critical urban development challenges such as climate change and urbanization and provides opportunities to create responses that address socio-ecological and politico-economic problems and inadequacies, thereby positioning cities to function effectively amidst turbulence (Cobbinah, 2021). Much of this research and the accompanying debates have been focused on Global North regions, countries, and cities. Across the Global South, discourse on urban resilience is evolving, particularly in some African, Chinese, and Indian cities. This has certainly been the subject of recent critique, such as its facilitation of disaster risk management (Anguelovski et al., 2018) flood management, and urbanization management (Andersen & Røe, 2016; Sager, 2014). These strategies improve resource-energy recycling, management and regeneration, and resource security, making them appropriate for resilience building (Meerow et al., 2016).

Framing DMA's (Saudi Arabia) urban resilience strategies within the broader Gulf Cooperation Council (GCC) context allows for a more nuanced understanding of regional policy responses to shared climate change and urbanization pressures. These countries face similar environmental and urban challenges, including extreme temperatures, water scarcity, rapid urbanization, and coastal exposure. For example, in Bahrain, resilience efforts have focused on mitigating coastal vulnerabilities (Almheiri et al., 2024) while Kuwait has explored the impacts of urban form expansion through agent-based modeling and GIS to better inform long-term planning and urban resilience strategies (Alghais & Pullar, 2015). Similarly, Oman's resilience discourse is increasingly shaped by its shift toward renewable energy governance and long-term sustainability planning (Sim, 2023), but in Qatar, holistic resilience frameworks have been adopted to manage rapid urban transformation and respond to compound socio-political and environmental shocks (Alraouf, 2021). Meanwhile, the United Arab Emirates has invested heavily in green infrastructure and pollution mitigation to reduce environmental risk across its coastal and desert cities (Farahat, 2016). These regional examples deepen the comparative perspective and position the DMA within a broader Gulf-wide discourse on urban sustainability and resilience. However, while these countries share regional challenges, the DMA presents a distinct case due to its oil-driven spatial development, low-density expansion, and infrastructure strain, factors that necessitate more tailored resilience strategies.

In response, this chapter answers two important research questions. (a) how do we engage with urban resilience to engender transformative strategies such as circular economy and nexus approach to planning (see Nhamo et al., 2023; Stevens, 2021) in regions that are less researched? (b) in what ways do the contours of rapid urbanization and climate change influence and diversify understandings of urban resilience via multiple adaptation pathways? On the first research question, less researched regions such as the Middle East have, by and large, been less directly involved with the global debates on urban resilience. Yet insights from such regions have the potential to open up and "diversify" understandings of urban resilience in ways that advance and consolidate urban sustainability in critical urban studies. As cities in the Middle East region rapidly urbanize and are exposed to the increasing risks of climate change (Aldhafeeri & Alhazmi, 2022; Mahmood et al., 2019), we argue that expanding research on urban resilience to key perspectives and transformative strategies in less researched regions beyond the ways that are often overlooked, can stimulate efforts toward urban sustainability.

We argue that circular economy and nexus planning are central to sustainability and resilience discourse (Nhamo et al., 2023; Stevens, 2021) operating within the urban nexus framework (see Sect. 2 for details). The urban nexus framework provides space for circular economy and nexus planning methods to improve resource-energy recycling, management and regeneration, and resource security, making them appropriate for sustainability planning (Nhamo & Ndlela, 2021; Schmidt, 2021). For example, nexus planning and the circular economy are essential components of urban nexus framework that guarantee the best possible utilization and handling of resources (Uddin et al., 2019; Almulhim & Al-Saidi, 2023). Therefore, as we argue, the urban nexus framework is an effective way to manage the mutuality of resources and provides the opportunity to assess how a proposal in one area affects the other associated sectors (Kim et al., 2015). However, unlike circular economy strategies, nexus planning has received limited research attention within urban nexus framework analysis mostly focusing on water and energy issues (see Sarikhani et al., 2023; Liang et al., 2022; Huang et al., 2023; Kolo et al., 2021; Xue et al., 2018). In response, this chapter demonstrates how the urban nexus framework remains central to addressing issues of urbanization via nexus planning producing innovative strategies toward building resilient cities in Saudi Arabia.

Profiling the DMA in Saudi Arabia, our aim in addressing the second research question is to show how the contours of rapid urbanization and climate change unpack and diversify understandings of urban resilience. We deconstruct the epistemological underpinnings of nexus planning within the urban nexus framework and then draw on recent scholarship in critical urban studies to provide entry points for ways of enhancing urban resilience and sustainability. This provides space to analyze how transformational adaptation strategies can be developed and applied in DMA, and the ways interdisciplinary and cross-sectoral collaborations can contribute to sustainable urban development. The chapter is organized as follows. Following the introduction, we provide an understanding and analysis of the urban nexus framework, emphasizing nexus planning which sets the context to explore pathways of urban resilience. Next, we present the methodology used in conducting this analysis. In the results section, we demonstrate the critical urban environmental challenges that threaten urban resilience efforts in DMA. We further develop an urban nexus framework and its relevance to building a resilient city. Finally, we present the discussion and conclusion sections.

8.2 The Urban Nexus Framework as an Urban Resilience Approach

To situate a diversified understanding of urban resilience and its implications on DMA, we identify insights from recent critical literature on urban nexus framework in urban planning and urban studies that are essentially helpful. The recent growth in the critical literature on urban nexus framework in the Global South, including its urban dynamic modeling possibilities, can help to unpack urban resilience in the Global South and provide fruitful ways toward sustainable urban development. The term "urban nexus" represents the relations, interconnectedness, and interdependencies in urban systems (e.g., energy, water, and food and material provisioning systems) and the role of integrated approaches in guiding transformational change (Lehmann, 2018). Although this section does not typically discuss urban resilience directly, we have here mobilized insights the urban nexus framework offers for diversifying urban resilience understandings. In a similar way, we apply the urban nexus framework to underscore the distinct and fundamental themes of urban resilience, providing critical contribution of diversified understandings to urban studies and planning.

Throughout discourses on urban resilience and sustainable development, the urban nexus framework has grown increasingly prevalent (Chen & Lu, 2015; Lehmann, 2018; Nhamo & Ndlela, 2021; Aquilina & Sheate, 2022; Koutra et al., 2022; Javan et al., 2023; Cervantes Puma et al., 2024; Giatti et al., 2024). This growth is underpinned by its capacity to integrate resource management processes that increase resource use efficiency and smart infrastructural systems, transform urban planning practice, and reduce gas emissions and waste generation. Central to the urban nexus framework is nexus planning which addresses the steps and approaches necessary for rendering the unified goal achievable by means of effective preparation and policy development. Together with circular economy, nexus planning provides an opportunity for accelerating urban sustainability and resilience by optimizing the security of resources, recycling, management, and regeneration within the frame of urban nexus thinking (Nhamo et al., 2020; Hettiarachchi & Ardakanian, 2016; Sperling & Berke, 2017). Nexus planning is a vital instrument for coordinating and structuring decision-making for a circular economy (Lehmann, 2018), revealing critical points for action which, in turn, drives scenario planning choices in urban development (Johnson & Karlberg, 2017).

Inherent in nexus planning portfolio are aspects of the sustainability transitions framework (Turnheim et al., 2015). This framework provides opportunity for nexus planning to integrate socio-technical transition theory, quantitative systems modeling, and real-world initiative-based learning insights into sustainable urban development. In Saudi Arabia where rapid urbanization, centralized planning, and increasing climate pressures demand transformative responses (i.e., a fundamental shift in approach), insights from the sustainability transitions framework become an enablement for nexus planning to manage these pressures. For example, the DMA, with its expansive infrastructure and exposure to ecological stress, provides a compelling case for applying such insights highlighting the urgency for long-term, multi-level governance strategies that address both institutional inertia and the complexity of systemic change. This will require an understanding of sustainability transitions and sustainable development interactions through scenario-based modeling (van Soest et al., 2019). O'Neill et al.'s (2014) shared socio-economic pathways framework adds to the nexus planning relevance in building urban resilience as it offers structured narratives for global development trajectories in the context of adaptation and mitigation challenges. While not specific to urban planning, the shared socio-economic pathways framework complements the urban nexus and circular economy approaches by providing a flexible basis for evaluating long-term, cross-sectoral interactions, such as those between water, energy, and land systems, and designing adaptive pathways for sustainable urban transformation under uncertainty.

The introduction of the Sustainable Development Goals (SDGs) nearly a decade ago further lends credence to the urban nexus framework by acknowledging the value of coordinated urban governance (Urbinatti et al., 2020), and the framework has continued to be incorporated into policymaking conversations over the years (Artioli et al., 2017) in cities worldwide, including Berlin and Kisumu (Heydkamp et al., 2014; Opiyo & Agong, 2020). The urban nexus framework's social aspects connected with justice and equality have also been emphasized in recent research (Jennings et al., 2019). Scholars such as Lehmann (2018) maintain that urban nexus framework should focus on variations in decision-making as well as accessibility to resources, particularly land, encouraging inclusive governance systems that allow those without an opportunity to have a greater impact. The function of technology in enhancing urban systems occupies the forefront in urban nexus investigations, despite Monstadt and Coutard (2019) cautioning people about technological determinism seeking an acceptable balance between equity and innovation. Overall, the idea of the urban nexus presents an insightful viewpoint for improving city resilience and supporting sustainable development in the face of rapid urbanization and climate change threats.

This chapter, therefore, applies the urban nexus framework to inform policy and decision-making on strategic policy formulations related to social, climatic, environmental changes and their impacts on urban landscapes for the DMA. The framework was developed by adapting insights from existing urban nexus and resilience planning literature (e.g., Aquilina & Sheate, 2022; Javan et al., 2023; Koutra et al., 2022; Nhamo et al., 2023) and was applied throughout the study to guide the structuring of data analysis, and interpretation of findings in alignment with the study's conceptual goals. The framework integrates the three main elements of sustainability, mainly ecological, economic, and social elements, holistically and provides a platform for transformational change that leads to sustainable cities. Within the context of Saudi Arabia's urban space, eight specific thematic areas of urban nexus framework are identified focusing on human-focused progress, sustainability, and resilience (see Fig. 8.1). They include intensive environmental and risk evaluation; sustainable investment and financing; governance and policy development; education and community involvement; nexus planning; technology and innovation; adaptive design and infrastructure; and the input, review, and execution systems. Within this

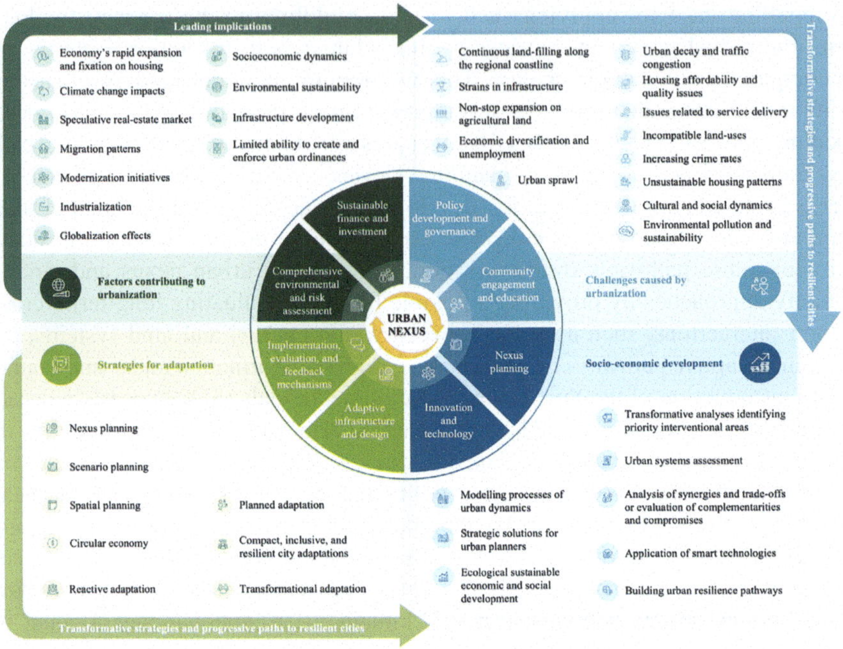

Fig. 8.1 A methodological framework for the urban nexus, illustrating the stages and directions toward resilient cities, while integrating the urban nexus approach with sustainability (Created by authors)

context, the urban nexus framework represents a planned, environmentally sound alternative for growth in metropolitan regions. It is within the foregoing context that this chapter is based.

8.3 Methods

8.3.1 DMA: Geography and Demography

DMA comprises the cities of Dammam, Al-Khobar, and Al-Dhahran, collectively forming a major urban region in the Eastern Province of Saudi Arabia. DMA lies in the eastern province of Saudi Arabia, at the Arabian Gulf coast between the waters of the Arabian Gulf and the dunes of Ad-Dahna desert (see Fig. 8.2). DMA is only ten meters above sea level and is situated in the Arabian Gulf in Eastern Saudi Arabia. It is located around 1230 km east of Jeddah and 400 km east of Riyadh, the country's capital. Over the last five decades, the region has grown to be an active center of trade, industry, and research, with a population of over 1.9 million

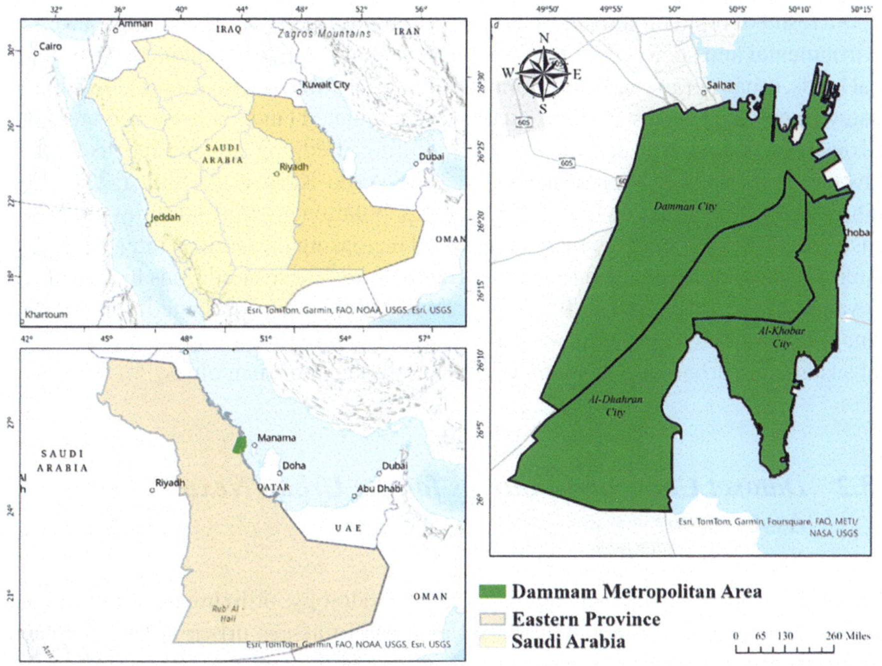

Fig. 8.2 Map showing the location of DMA in Saudi Arabia. *Source* authors

(AbouKorin et al., 2020). Oil was discovered in 1938, and DMA experienced an extensive transformation. Between 1940 and 1950s, additional oil deposits were discovered across the metropolis. These oil fields made up 25% of the globe's found oil reserves, which led to the development and rise in infrastructure building (UN-Habitat, 2019), propelling the city to become the Eastern Region's capital and a major global center for oil shipment.

DMA has faced a relatively significant urban expansion, with its population rising from 365,000 in 1974 to 1.9 million 2020, with projections of 3.62 million by 2040 (AbouKorin et al., 2020). The city now features a consolidated urban structure because of its fast expansion. Most of this expansion occurred outside of the MoMRA-established urban boundaries and needed to be more anticipated. For most regular individuals, making home investments has become a favored option. The housing sector is seeing significant investment from private-sector enterprises that have employed a similar strategy (Abou-Korin & Al-Shihri, 2015). Inside the urban growth limit, 8900 of the 25,618 hectares, around 35%, of land had been built upon in 2003, keeping 16,700 hectares of undeveloped land (Al-Shihri, 2016). DMA's explosive economic growth, the opportunistic real estate market, and the Eastern Province Urban Planning Department's insufficient ability to create and enforce urban laws represent the three primary drivers for the city's significant expansion.

As a result of rapid urbanization, DMA faces multiple challenges creating a web of environmental and socio-economic concerns, such as immediate damage of agricultural lands, urban sprawl, and deterioration of land and marine environment leading to wastewater control issues, poor air quality, contaminated land, increased congestion, and destruction of coastal habitats due to landfilling, dredging, and marine discharges (Alwadei et al., 2022; Abdelrahman et al., 2021; Abou-Korin & Al-Shihri, 2015; Al-Shihri, 2016). This situation is fueled by rapid population growth and sprawling, low-density urban development, which have placed increasing pressure on interconnected systems of water, energy, land use, and infrastructure. These conditions highlight the relevance of the urban nexus framework, which offers a coordinated approach to managing these interdependencies and is critically important for advancing resilient and sustainable urban development in DMA and other Arabian cities.

8.3.2 Dataset Collection Sources for the Urban Nexus Assessment

This study adopts a quantitative geospatial methodology, utilizing satellite-derived environmental indicators and GIS-based analysis to assess urban nexus. For land use and land cover (LULC) mapping, we used LULC data derived from Landsat images for the years 1993, 2003, 2013, and 2023. Two of these images were from the Landsat 5 TM sensor, and the other two have a Landsat 8 OLI sensor with a spatial resolution of 30 m. The cloud coverage was < 10%. For radiometric correction, we used the Radiometric Correction Tool in ERDAS Imagine, which helped convert digital numbers into radiance. This helped to standardize the brightness values by compensating for sensor-specific variances and correcting for sensor noise and artifacts. Following this, atmospheric correction was performed to convert radiance to surface reflectance, effectively minimizing atmospheric interference such as water vapor, aerosols, and other particulates. For this, we applied the ATCOR (Atmospheric and Topographic Correction) method. For mapping the air pollutants, we used Sentinel-5P satellite image data from the TROPOMI sensor. As TROPOMI data is available from 2018, we chose 2018 and 2023 data. Table 8.1 shows the details of these datasets.

Table 8.1 Details of the dataset

Year	Satellite	Sensor	Resolution	Cloud cover	Bands
1993, 2003	Landsat	Landsat 5 TM	30 m	< 10	1 - 7
2013, 2023	Landsat	Landsat 8 OLI	30 m	< 10	2 – 7, 10, 11
2018, 2023	Sentinel-5P	TROPOMI	$NO_2 = 3.5$ km × 7 km $SO_2 = 7$ km × 7 km $CO = 7$ km × 7 km $CH_4 = 7$ km × 7 km		

Table 8.2 Description of classified LULC

LULC classes	Description
Built-up area	Urban settlements, buildings, impervious surface, roads
Waterbody	Waterbodies, wetlands, rivers, ponds, and canals
Vegetation	Vegetation, forest land, and trees
Barren soil	Desert, sand, and rocks
Cropland	Agriculture land, grassland

8.3.3 LULC Mapping

This study considered five types of land cover classes, including built-up area, vegetation, waterbody, cropland, and barren soil areas to assess the spatiotemporal changes in LULC classes driven by urbanization in the study region. In this regard, we used LULC data derived from Landsat images (Landsat 5 TM for the years 1993 and 2003, Landsat 8 OLI for 2013 and 2023) by applying the random forest classifier method. RF is an ensemble learning method that builds multiple decision trees and merges their outputs to improve LULC classification accuracy (Amini et al., 2022). Each decision tree was trained on a random subset of the data with a random selection of features, which helps reduce overfitting and increases the generalizability of the model. RF can handle high-dimensional datasets and performs better in terms of noise and outliers in the data (Nguyen et al., 2018; Rodriguez-Galiano et al., 2012). To train the dataset, we selected 75 to 100 sample points for each of the LULC classes. The readers are referred to Shih et al (2022) and Tikuye et al. (2023) for better insights of the RF classifier. Based on the derived LULC, we have classified the LULC of DMA into five classes (Table 8.2).

8.3.4 Measurement of Land Surface Temperature

Land Surface Temperature (LST) is the measure of the temperature of the Earth's surface sensed by satellite images. Unlike air temperature, which is measured at weather stations a few meters above the ground, LST represents the heat emitted by the surface itself, including soil, vegetation, urban areas, and water bodies. LST is derived from thermal infrared (TIR) data captured by satellite sensors such as Landsat, MODIS, and ASTER (Hulley et al., 2019; Rajeshwari & Mani, 2014). In this study, we calculated the LST for the years 1993, 2003, 2013, and 2023. For this, we used the Landsat 5 TM and Landsat 8 OLI images. The process of calculating LST from satellite imagery entails processing thermal infrared (TIR) data to obtain temperature values. In the following section, we detailed the steps for calculating LST from Landsat 5 TM and Landsat 8 OLI images collected in this study.

8.3.4.1 Radiometric Calibration

For radiometric calibration of Landsat 5 TM, first digital number (DN) of the thermal band (band 6) to spectral radiance using Eq. 8.1 (Li et al., 2013; Sobrino et al., 2004):

$$L_\lambda = \frac{(L_{max} - L_{min}) \times (DN - Q_{calmin})}{Q_{calmax} - Q_{calmin}} + L_{min} \tag{8.1}$$

Here, L_λ = spectral radiance, L_{max} and L_{min} area the radiance scaling factors, Q_{calmax} and Q_{calmin} are calibration constants.

For Landsat 5 TM, we converted the digital number (DN) values of the thermal band (Band 6) to spectral radiance using Eq. 8.2 (Rajeshwari & Mani, 2014).

$$L_\lambda = (ML \times DN) + AL \tag{8.2}$$

Here, ML is the multiplicative rescaling factor, AL is the additive rescaling factor.

8.3.4.2 Brightness Temperature (TB) Calculation

We converted spectral radiance to brightness temperature (in Kelvin) for both Landsat 5 and Landsat 8 using the inverse Planck function in Eq. 8.3 (Sresto et al., 2022; Fattah et al., 2021).

$$T_B = \frac{K_2}{\ln\left(\frac{K_1}{L_\lambda} + 1\right)} \tag{8.3}$$

where K_1 and K_2 are the thermal constants specific to each Landsat sensor.

8.3.4.3 NDVI Calculation

We calculated the Normalized Difference Vegetation Index (NDVI) for each image using the near-infrared (NIR) and red bands. For this we applied Eq. 8.4 (Gascon et al., 2016; Yengoh et al., 2015).

$$NDVI = \frac{NIR - Red}{NIR + Red} \tag{8.4}$$

8.3.4.4 Land Surface Emissivity (LSE)

In this stage, we calculated LSE ($\in$) based on NDVI value using Eq. 8.5 (Sresto et al., 2022; Fattah, Morshed, & Morshed, 2021).

$$\in = 0.004 \times \left\{ \left(\frac{NDVI - NDVI_{min}}{NDVI_{max} - NDVI_{min}} \right)^2 \right\} + 0.986 \quad (8.5)$$

8.3.4.5 LST Calculation

Using Eq. 8.6, the TB values were converted to LST (in Kelvin) applying emissivity correction.

$$LST = \frac{T_B}{1 + \left(\frac{\lambda \times T_B}{\rho} \right) \ln \in} \quad (8.6)$$

Here, λ is the wavelength of emitted radiance (11.5 μm for Landsat 5 TM and 10.9 μm for Landsat 8 Band 10 and 12.0 μm for Band 11), and $\rho = 1.438 \times 10^{-2}$ mK. We have converted the LST from Kelvin to °C by subtracting 273.15 (De Jesus & Santana, 2017; Jimenez-Munoz et al., 2014).

8.4 Results: Urban Resilience Analysis in DMA

8.4.1 Land Use Land Cover (LULC) Change Dynamics

Figure 8.3 presents the spatial distribution of LULC classes in DMA over the past four decades and findings. Table 8.3 indicates that barren soil areas were the dominant LULC type in the study region. In 1993, barren soil covered approximately 81.50% (1162.59 km^2) of the study area, primarily distributed across the western region. Settlements or built-up areas were found across 201.97 km^2, predominantly distributed in the eastern and northeastern regions. During the period 1993–2003, built-up areas expanded from 201.97 km^2 (14.16%) to 419.17 km^2 (29.39%), with expansion toward the western side of the study area. Over the entire study period, built-up areas expanded by 217.20 km^2, with the highest expansion (77.08 km^2) occurring between 2013 and 2023. Conversely, the total loss of barren soil areas were calculated to be 174.43 km^2, with the highest decline (64.91 km^2) occurred during 2013–2023. Figure 8.4 shows that expansion of urban areas and declination in barren soil areas occurred in the same direction. DMA experienced significant decline in barren soil areas.

Interestingly, vegetated areas expanded by 5.02 km^2 in the first decade, but then declined in subsequent decades. Urban expansion has had the most drastic impact on waterbodies. Despite being a desert region, approximately 32.91 km^2 of waterbodies were transformed into settlements, especially along the coastal areas of the cities. These changes support the findings of Rahman et al (2017) who reported increase in

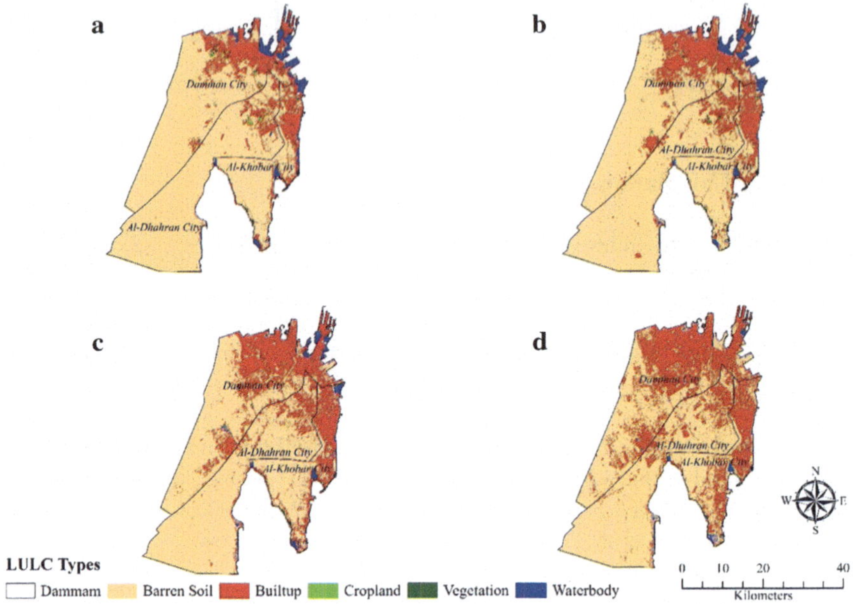

Fig. 8.3 Spatial distribution of LULC classes in DMA in **a** 1993, **b** 2003, **c** 2013, and **d** 2023

Table 8.3 LULC statistics in DMA during 1993–2013

Class	Area	1993	2003	2013	2023
Built-up area	%	14.159%	18.841%	23.982%	29.385%
	km^2	201.97	268.76	342.09	419.17
Vegetation	%	0.523%	0.874%	0.387%	0.346%
	km^2	7.45	12.47	5.52	4.93
Waterbody	%	3.274%	2.701%	1.779%	0.967%
	km^2	46.71	38.52	25.37	13.80
Cropland	%	0.543%	0.066%	0.029%	0.028%
	km^2	7.74	0.94	0.41	0.402
Barren soil	%	81.502%	77.519%	73.824%	69.274%
	km^2	1162.59	1105.78	1053.07	988.16

urban areas and decrease in barren soil, vegetation, and waterbody areas in DMA. Moreover, rapid urban area expansion in major Saudi Arabian cities reported in several prior studies including (Alqurashi & Kumar, 2014; Alsharif et al., 2022; Mohammad et al., 2022). Consistent with this study, several studies have reported the increase in vegetation cover areas during 1990 to 2000 (Aina et al., 2008).

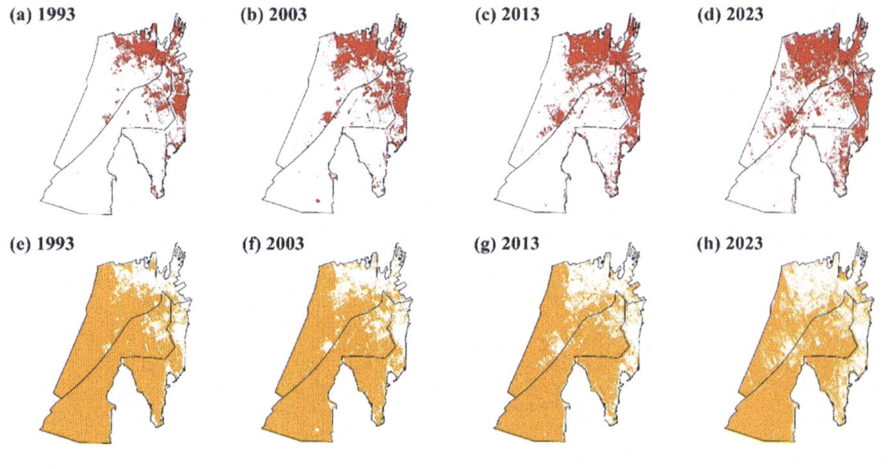

Fig. 8.4 Spatial extent of built-up area **a–d** and barren soil area change **e–h** in DMA during 1993–2023

8.4.2 Urbanization Pressures on Natural Ecosystems in DMA

Figure 8.5 shows the spatial distribution of changes in natural ecosystems in DMA between 1993 and 2023. Urbanization in DMA has led to a decline in natural ecosystems, such as vegetated ecosystems, aquatic ecosystems, and agricultural ecosystems. Figure 8.5 shows that DMA initially had a total of 32.91 km^2 of aquatic ecosystems, primarily located on the northern and northeastern sides of the study area. The decline of the aquatic ecosystem area was most prevalent in Dammam City, followed by Al-Khobar City and Al-Dhahran City (Fig. 8.6). The results suggest that approximately 22.75 km^2 of the aquatic ecosystem area transformed into barren soil cover during 1993–2023, with another 13.98 km^2 transitioning to urban settlements (Table 8.4). Only 9.30 km^2 of the original aquatic ecosystem area remained untransformed. AbouKorin et al. (2020) reported that population growth has been accelerating urbanization in the DMA, further impacting the coastal ecosystem. Figure 8.6 shows the transformation of aquatic or coastal ecosystems into urban settlements in the DMA. Aldegheishem (2024) investigated the major drivers of ecological changes in Saudi Arabia and found that urbanization is the leading cause of ecological degradation, which supports the findings of this study. Several prior studies claimed that such ecological degradation has led to environmental degradation, greenhouse gas emissions, and biodiversity loss (Amer et al., 2022; Crippa et al., 2023; Raggad, 2018). To conserve natural ecosystems, governments have been focusing on preserving them and promoting research to increase awareness and policy adaptation (Alatawi, 2022).

The second most affected ecosystem is the agricultural ecosystem, which declined by 7.34 km^2. Table 8.4 shows that built-up areas expanded across approximately 2.85 km^2 of agricultural ecosystems in DMA during 1993–2023. Dammam City experienced the highest loss, followed by Al-Dhahran City and Al-Khobar City. Barren soil

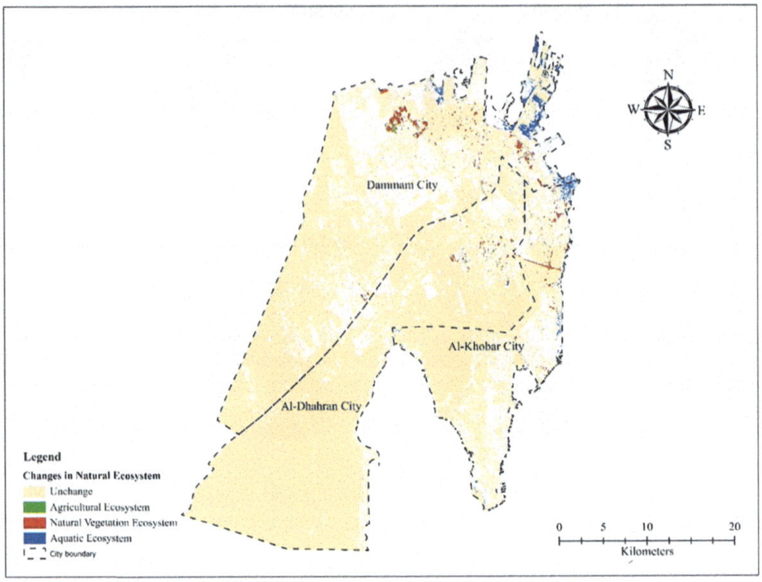

Fig. 8.5 Transformation of natural ecosystems into urban areas in DMA

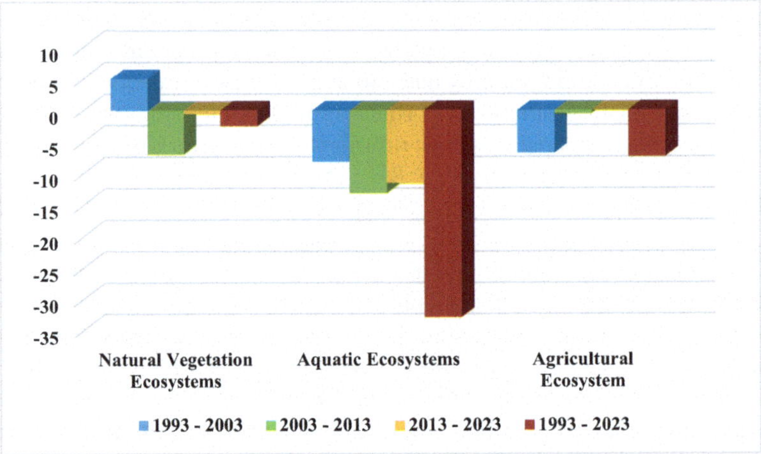

Fig. 8.6 Changes in natural ecosystems in DMA during 1993–2023

area expanded over 4.11 km^2 of agricultural ecosystems. Vegetated ecosystems experienced a lower degree of transformation, totaling 7.23 km^2 (Table 8.4). About 3.45 km^2 of the vegetated ecosystem area transformed into barren soil covered areas, and 3.15 km^2 into urban settlements. The transition matrix revealed that approximately 2.01 km^2 of barren soil areas, 0.35 km^2 of agricultural ecosystem area, and 0.20 km^2

Table 8.4 Tranformation of natural ecosystems during 1993–2023

LULC Class	Barren soil	Built-up areas	Cropland	Vegetation	Waterbody	Total
Agricultural Ecosystem	4.11	2.85	0.02	0.35	0.02	7.34
Vegetated Ecosystem	3.45	3.15	0.01	0.57	0.05	7.23
Aquatic Ecosystem	22.75	13.98	0.00	0.20	9.30	46.23

of aquatic ecosystem area transformed into vegetated ecosystems. However, only 0.57 km^2 of the natural vegetated ecosystem area remained unchanged. Figure 8.5 shows that the transformation of vegetated ecosystems occurred primarily across the northern and northeastern regions of the study area, most prominently in Dammam. As a result of declining natural vegetated ecosystems, the city may face increased air pollution and higher temperatures due to the decline in natural cooling and air purifying functions that plants provide. This further exacerbates the urban heat island effect. The decline in Dammam's aquatic ecosystems can result in compromised water quality, leading to a decline in aquatic biodiversity and impacting water availability for drinking, agriculture, and industrial uses. Climate change may potentially compound the adverse impacts by altering precipitation patterns, increasing ambient and surface temperatures, and triggering extreme weather events, which can lead to more frequent heatwaves, droughts, or flooding in the study area. The loss of these ecosystems also diminishes the city's resilience against natural disasters; for example, wetlands and floodplains play critical roles in natural water filtration and flood mitigation (Amenu & Mamo, 2018; Fluet-Chouinard et al., 2023). Socially, urban residents may experience a lower quality of life due to the reduced aesthetic, recreational, and psychological benefits of natural environments (Shirazi & Kazmi, 2016). Economically, the consequences include increased healthcare costs due to poorer air quality and heat-related illnesses, higher expenditures on artificial cooling systems, and potential losses in tourism revenue in cities known for their natural landscapes (Deng et al., 2013; Hasan et al., 2020; Hussain et al., 2023).

8.4.3 Spatial LST Distribution During 1993–2023

Figure 8.7 presents the spatiotemporal distribution of LST in DMA during 1993–2023 and the key findings are summarized in Table 8.6. We classified the LST range for all four study years into five classes. Figure 8.8 shows the changes in these classes during the study period, and Table 8.5 suggests the rise in mean, maximum, and minimum LST across all three cities. In DMA, the mean LST has risen from 32.37 °C in 1993 to 38.19 °C in 2023, indicating a significant increase in average temperatures. The maximum LST values show a notable increase, which is particularly significant in DMA, where it increased from 37.61 °C in 1993 to 46.09 °C in 2023.

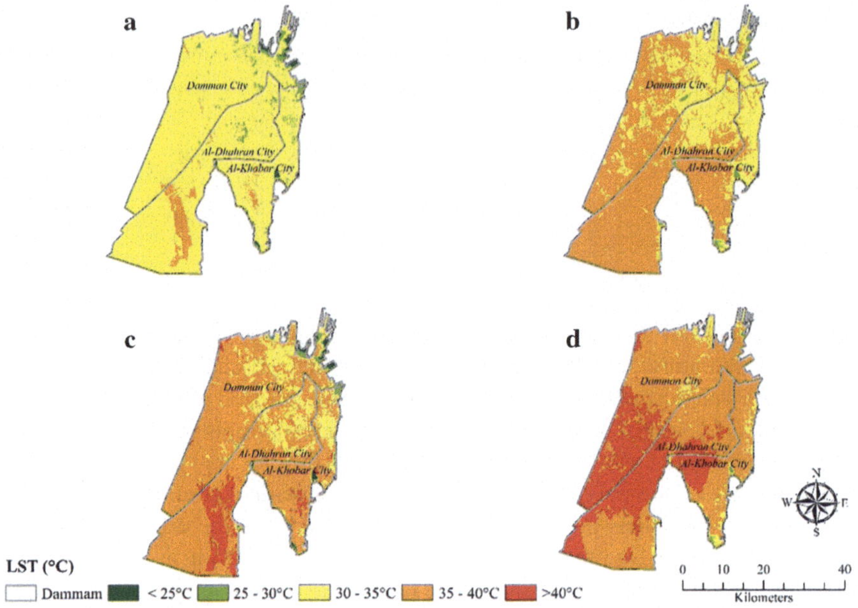

Fig. 8.7 Spatial distribution of LST patterns in DMA in **a** 1993, **b** 2003, **c** 2013, and **d** 2023

The standard deviation values suggest variations, but they generally trend upward, implying increasing variability in surface temperatures. DMA's STD increased from 1.94 in 1993 to 2.39 in 2023, and Al-Khobar saw an increase from 2.15 to 2.61. This increased variability suggests that the disparity of heat experienced across different areas has been increasing, and the study area is experiencing more frequent extreme temperature events, contributing to more pronounced heatwaves across all the cities. This upward trend in both mean and maximum temperatures indicates significant warming in the study region.

Table 8.6 shows that in 1993, about 1.37% (19.60 km^2) of the study area's surface temperature was recorded below 25 °C, distributed across the coastal side of DMA and the southern side of Al-Khobar City (Fig. 8.8q). In the following decade, the area with LST below 25 °C decreased to only 0.04 km^2 (Fig. 8.8s). However, by 2023, this area had increased again to 16.69 km^2 (1.17%) (Fig. 8.8s). The dominant LST class across the study region ranged from 30 to 35 °C, accounting for approximately 89.66% of the total area in 1993. Figures 8.8i-1 show the decline of these areas over time, starting from the western region. In the first decade, the area with LST between 30 and 35 °C declined from 89.66% to 32.61%. By 2023, only 5.46% (77.92 km^2) of the total area fell within this LST range, scattered across the eastern region (Fig. 8.8h).

In contrast, areas with higher LST values (35–40 °C and > 40 °C) showed an increasing trend throughout the study period. Notably, between the first three decades (Fig. 8.8e-g), areas with LST of 35–40 °C increased across the western region,

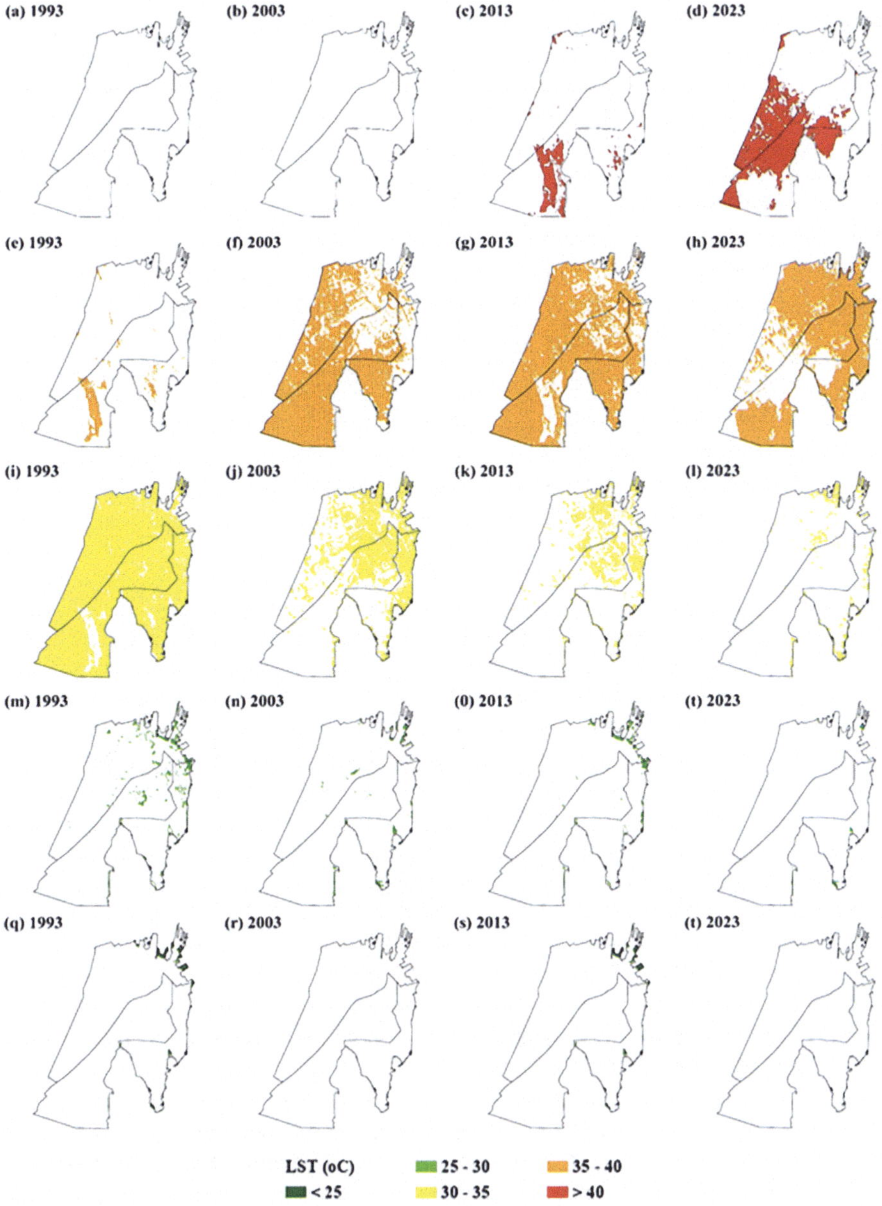

Fig. 8.8 Spatial changes in different LST classes during 1993–2023

Table 8.5 Recorded LST statistics across different cities in DMA

Year	City	LST (°C)				
		Mean	Max	Min	Range	STD
1993	Dammam	32.37	37.61	23.25	14.36	1.94
	Al-Khobar	32.55	36.44	23.25	13.19	2.15
	Al-Dhahran	33.16	38.39	24.11	14.27	1.51
2003	Dammam	35.06	41.24	25.56	15.68	1.58
	Al-Khobar	35.14	39.64	25.59	14.05	2.08
	Al-Dhahran	35.47	39.66	24.66	15.00	1.51
2013	Dammam	36.05	42.96	23.25	19.71	2.84
	Al-Khobar	36.58	41.45	23.25	18.20	3.15
	Al-Dhahran	37.71	43.71	23.68	20.03	2.29
2023	Dammam	38.19	46.09	26.16	19.93	2.39
	Al-Khobar	37.18	43.08	27.09	15.99	2.61
	Al-Dhahran	39.26	45.01	27.53	17.48	1.97

Table 8.6 Spatial LST Distribution in DMA during 1993–2023

LST range (°C)	Area				
		1993	2003	2013	2023
< 25	%	1.37%	0.00%	1.17%	0.00%
	km^2	19.60	0.04	16.69	0.00
25 – 30	%	4.10%	1.52%	1.93%	0.50%
	km^2	58.48	21.67	27.58	7.08
30 – 35	%	89.66%	32.61%	15.77%	5.46%
	km^2	1279.00	465.20	224.90	77.92
35 – 40	%	4.86%	65.84%	74.52%	64.19%
	km^2	69.38	939.20	1063.00	915.70
> 40	%	0.00%	0.00%	6.58%	29.82%
	km^2	0.00	0.06	93.87	425.40

which previously belonged to the 30–35 °C LST class. Between 1993 and 2013, the area with LST of 35–40 °C increased dramatically, from 4.86% (49.38 km^2) to 74.52% (1063 km^2) of the study region. However, in the final decade, this class area declined to 64.19% (915.70 km^2) of the total study area, with changes occurring across the southwestern region of Dammam, northeastern Al-Khobar City, and the central region of Al-Dhahran City (Fig. 8.8h). Notably, Fig. 8.8d shows that these regions experienced a further rise in LST, exceeding 40 °C. While areas with LST above 40 °C were absent in 1993 and 2003, this percentage increased to 6.58% in 2013. Finally, in 2023, approximately 425.40 km^2 of the study area exhibited LST exceeding 40 °C, indicating a substantial rise in LST and higher potential of heat island effects. The LST change in this study is consistent with Rahman et al. (2017). Several prior studies reported the drastic rise in surface temperature across different regions of Saudi Arabia (Ahmed et al., 2024; Safieddine et al., 2022; Sekertekin & Zadbagher, 2021), and these studies claimed that the cities of KSA would experience substantial heat island effects.

8.4.4 Influence of Natural Ecosystem Loss on LST in DMA

Figure 8.9 presents the recorded LST across different LULC classes, while Fig. 8.10 shows the changes in LST over the study period. Figure 8.9 shows that aquatic ecosystems consistently recorded the lowest LST throughout the study years. However, as shown in Fig. 8.10b, LST in these ecosystems increased over time. The mean LST for aquatic ecosystems rose from 26.72 °C in 1993 to 32.56 °C in 2003, then to 34.08 °C in 2013, before decreasing slightly to 32.67 °C in 2023. Notably, the standard deviation value suggests significant variations in LST across aquatic ecosystems, except perhaps in 1993. Vegetated ecosystems exhibited the second-lowest mean LST, with values of 30.21 °C, 41.39 °C, 43.02 °C, and 34.38 °C in 1993, 2003, 2013, and 2023, respectively. Interestingly, vegetated areas consistently recorded the lowest maximum LST across all ecosystem types during the study period, highlighting their potential for mitigating thermal stress. Previous studies focusing on different regions of Saudi Arabia have reported lower LST across vegetated areas (Ahmed et al., 2024; Mallick et al., 2022; Safieddine et al., 2022). Addas (2023) reported that declination in vegetated areas led to elevated LST in Jeddah and Alshehri et al. (2023) in Riyadh. These findings underscore the importance of vegetation in mitigating urban heat islands and regulating local climates in arid regions like Saudi Arabia.

Similar to vegetated areas, and cropland areas generally showed lower LST compared to barren soil and urban areas. Mean LST across cropland areas ranged from 30.78 °C in 1993 to 42.01 °C in 2013. In contrast, barren soil areas exhibited a range of 33.28 °C to 44.54 °C, and built-up areas ranged from 31.47 °C to 43.16 °C. These findings suggest that urban and barren soil areas experienced higher LST and have a lower capacity for managing thermal conditions. Rahman et al. (2017) investigated the LST in DMA in the year 1990, 2002, and 2014. They recorded higher LST across barren soil cover areas, followed by built-up areas which is consistent with the findings of this study. Overall our analysis suggests that loss of natural ecosystem areas has been leading to higher LST in the study areas. However the findings in Fig. 8.10 suggest that other factors may influencing this increase.

8.4.5 Analysis of Air Pollutant Concentration in DMA

Analysis of air pollutant concentration was conducted to understand the extent of vulnerability of DMA. Figures 8.11a, b show a rise in NO_2 concentration levels

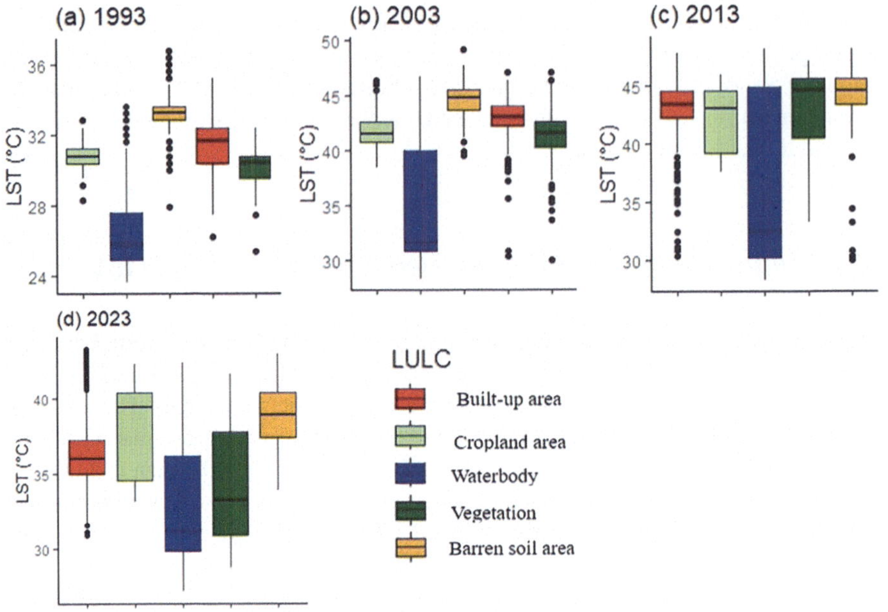

Fig. 8.9 Distribution of LST over different LULC classes

with the highest recorded NO_2 concentration increasing from 0.0001815 mol/m^3 to 0.00205 mol/m^3. Notably, even the lower concentration levels of NO_2 also increased. Figures 8.11c, d depict the rise in SO_2 concentration levels, particularly across DMA. The range of CH_4 concentration increased from 1858.63 ppb to 1933.17 ppb in 2018 to 1885.48 ppb to 1938.59 ppb in 2023, with the highest increase observed in DMA. However, CH_4 concentration levels in the central region of Al-Khobar City have declined.

The concentration range of CO in the study region also increased. In 2018, CO levels ranged from 0.0315768 mol/m^3 to 0.034552 mol/m^3. This range increased to 0.031889 mol/m^3 to 0.0368375 mol/m^3 in 2023. The loss of vegetated and aquatic ecosystems, which have the potential to purify air, likely explains this rise in CO concentration. Labban and Butt (2023) identified the Dammam as having high aerosol concentrations and reported an increase in these levels, which aligns with the changes in air pollutant concentrations observed in this study. Similarly, Hassan et al. (2022) noted a rise in major air pollutants (NO_2, CO, SO_2) in Riyadh during 2019–2020, while Rehan and Munir (2022) observed increased NO_2 and O_3 concentrations in Jeddah during 2018–2019. Additional studies have documented the rise of air particles across various regions of Saudi Arabia (e.g., Hussein et al., 2014; Porter et al., 2014; Alharbi et al., 2015; Alghamdi, 2016; Harrison et al., 2017; Lim et al., 2018). Alwadei (2022) attributed the increasing concentrations of $PM_{2.5}$ in Dammam to dust storms and rapid urbanization. These findings collectively highlight a concerning trend of escalating air pollution in DMA, underscoring the need for effective air

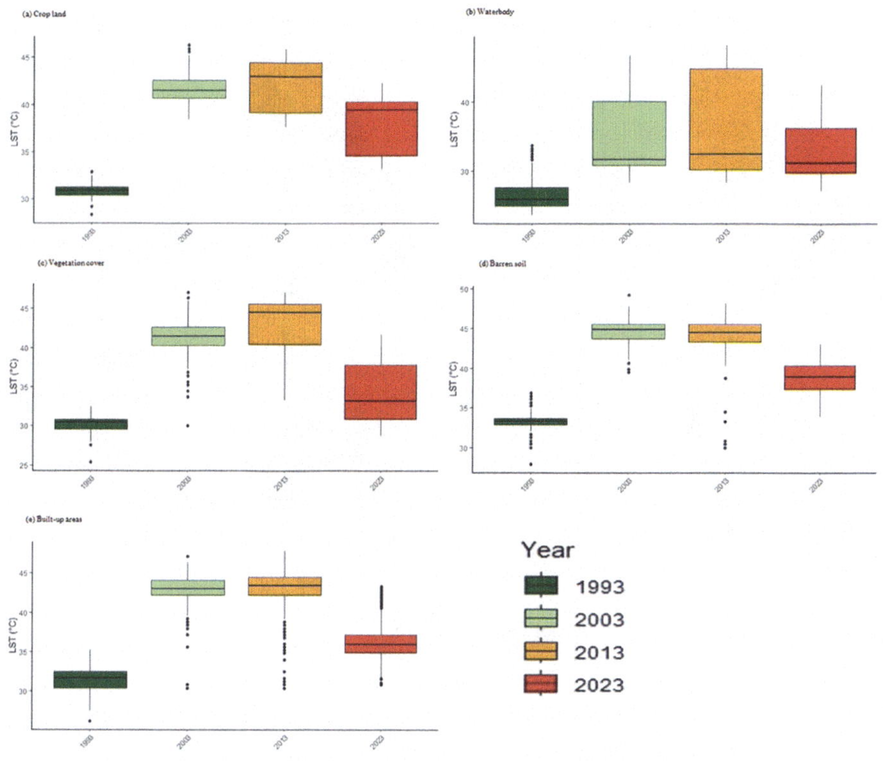

Fig. 8.10 Changes in LST distribution over different LULC classes

quality management and mitigation strategies to address the health and environmental impacts of rising pollutant levels.

8.5 Contours of Urban Resilience in DMA: Urban Nexus Positionality

Our analysis shows that the DMA is currently affronted by cross-sectional, diverse, and interconnected challenges. This situation calls for revolutionary strategies that provide impartial, comprehensive analysis toward resilient futures. Saudi Arabia's Vision 2030, which was approved as a development plan for the Kingdom of Saudi Arabia, supports this drive for economic diversification, following its objectives of promoting technological advancement and competitiveness in various areas and facilitating the growth of the economy in new industries (Vision 2030, 2018). One crucial method for addressing the geographical features of Vision 2030 is the National Spatial Strategy 2030, which recognizes that to achieve a sustainable and resilient

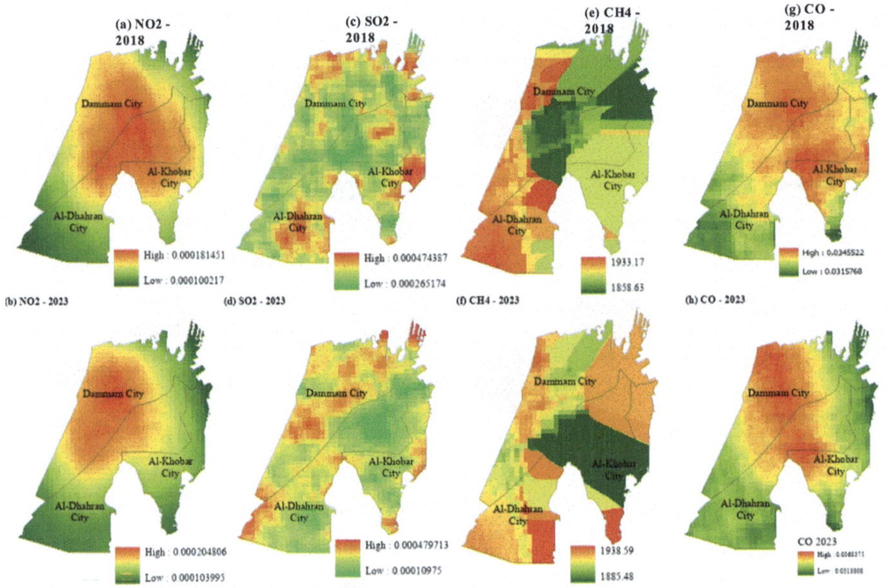

Fig. 8.11 Spatial distribution of air pollutants in DMA

Kingdom of Saudi Arabia, a balanced spatial strategy is required. Nonetheless, the National Spatial Strategy 2030 (NNS) and national development plans must be revised, and the strategy must be connected to a national planning system to close regional development gaps (UNDP, 2021). Thus, by encouraging waste reduction throughout the whole value chain of the economy, we argue for progressive revolutionary routes to direct the design and development of a balanced urban system with unified resource management (Yang et al., 2023). An essential part of both urban planning and sustainable resource management is the suggested urban nexus framework. It covers the vast connections and interdependence across urban systems, especially those for ensuring the supply of food, water, energy, and materials. Nexus planning, then, offers the framework for combining diverse but related industries into a cohesive strategy where all components work together effectively to realize a set result for the group. Guaranteeing effective utilization of resources, sustainable infrastructure systems, innovative planning techniques, waste management, and carbon emission reductions are the cornerstones of the urban nexus framework (Fig. 8.12). It offers a mechanism to put the provincial framework for spatial infrastructure into practice.

The findings presented here are directly informed by the methodological framework shown in Fig. 8.1, which guided the study's analysis and thematic areas of the urban nexus. Figure 8.12 presents a contextualized urban nexus framework for DMA and highlights the procedures involved in implementing measures to enhance resilience and manage fast urbanization. Policymakers are guided by the adaptation processes in developing programs that encourage resilience. This integrated evaluation of urban systems links the drivers to urbanization, urban metabolism, economic

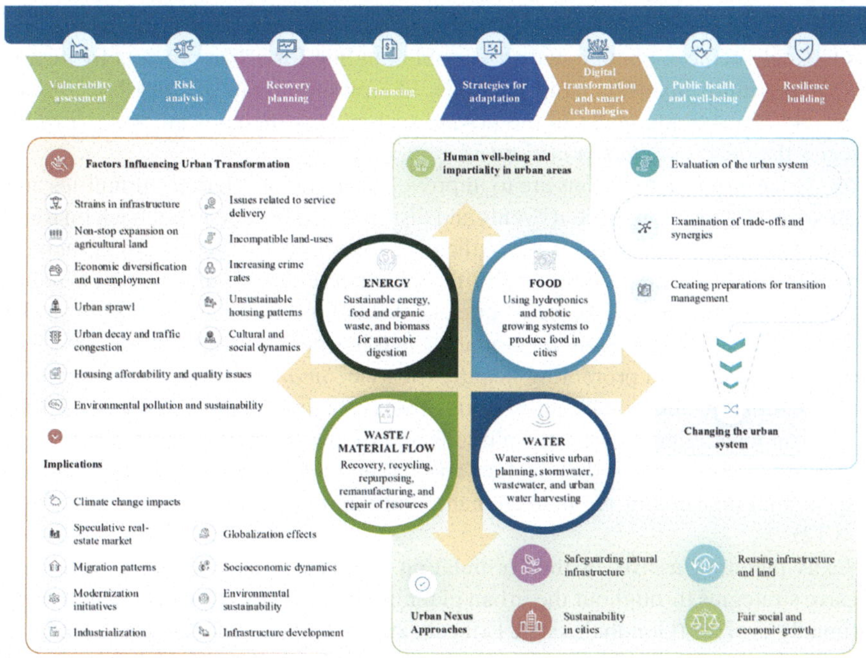

Fig. 8.12 A framework for DMA's contextualized urban nexus that illustrates the steps involved in transformational urban planning (Created by authors)

pragmatic principles, environment and resources, governance of urban areas, sustainable development, security and geopolitics, and other urban input streams (Chen & Lu, 2015; Emamjomehzadeh et al., 2023). Employing scenario planning, urban nexus modeling gives information on risks, trade-offs, and synergies, as well as options for resolution (Lehmann, 2018; Canyon, 2021). For example, smart technology, involving devices that use artificial intelligence, internet of things, and automation to optimize performance, increase efficiency, and enhance user experience, can be used to boost resilience and adaptation, and financial resources can be allocated to enable efficient deployment and monitoring (Almulhim, 2025; Argyroudis et al., 2022).

To be able to further understand these concepts, a drawn upon *eight variable* themes were specified to be able to attain urban sustainability. In DMA, *vulnerability assessment* is critical for resilience building and revolves around comprehending and picking the vulnerabilities present within urban governance, infrastructure, and social fabric, causing urban centers more vulnerable against economic, social, and environmental difficulties. This provides space for *risk analysis* which involves assessing the potential risks confronted by urban areas, like social and economic instabilities, repercussions from climate change, and natural disasters, and aims to forecast future planning and risks for unexpected cases. In an increasingly climate vulnerability city-region, *recovery planning* remains another resilience consideration, revolving around developing strategies for rapid and efficient recovery following disasters, and aims for restoring vital infrastructure, support, and services. *Financing* functions to safely achieve enough and sustainable funding for urban projects by means of exploring innovative financing methods, like public-private partnerships, green bonds, and investments made in the private sector. This includes tailoring specific strategies that lead *adaptation initiatives*.

Resilience building functions are to improve the ability of urban communities and systems to withstand unexpected events and disasters and be able to get back on track after such situation by means of strengthening government systems, infrastructure, and community networks. Resilience efforts include *public health and well-being* prioritizing health and well-being of citizens through urban design and policy, like making healthcare services more available, making public spaces more open and safer for all people, and promoting healthy lifestyle. *Smart technologies and digital transformation* are additional opportunities for leveraging urban resilience in DMA focusing on leveraging smart city solutions and digital technology to further boost urban sustainability, via utilizing data and technology to achieve greater citizen engagement, enhanced urban management, and improved delivery of services (see Fig. 8.12).

Nexus planning assists in the formulation of policies and decisions to create cohesive strategies throughout the urban planning process since the urban landscape continually evolves (Gondhalekar & Erlbeck, 2021). Unplanned changes to land use influence critical infrastructure and put the health of people and the environment in danger (Su et al., 2021). To improve institutional performance, management of resources, and quality of service, the nexus planning helps stakeholders find and explore potential synergies across sectors, governments, and technical disciplines. It challenges the trade-offs, fragmented duties, and traditional sectoral mentality that frequently leads to uncoordinated investments, higher costs, and underused infrastructures and amenities (Lehmann, 2018). The ultimate objective of the urban nexus conceptual framework is to achieve sustainable manufacturing, urban resilience, and an urban circular economy in urban settings, overall accelerating access to services and improving the quality of life of the world's inhabitants (UN-ESCAP, 2016). The key is striking an acceptable compromise between economic growth and the preservation of the environment and natural resources, with a focus on resource efficiency and recycling as well as safeguarding the environment (Ghisellini et al., 2016). The integration of sustainable manufacturing processes, the establishment of eco-industrial parks, and the incorporation of resource-based planning for development in sectors such as agriculture, industry, and urban areas distinguish it as a linear economy strategy or follows a make-use-dispose strategy from which it is characterized by minimal utilization of energy and production of harmful substances as well as substantial effectiveness in utilization of resources (Ghisellini et al., 2016). In contrast, the urban nexus approach considers the effects of fast urbanization, population growth, and climate change. It fosters the reuse of already-developed land, involving the rehabilitation of deteriorated infrastructure while reducing the amount of new greenfield development (Nhamo & Ndlela, 2021; Nhamo et al., 2023).

8.6 Discussion

8.6.1 Scholarly Insights

Human geography and urban studies literature predisposes us to be more attuned to how urban resilience is understood, conceptualized, and contextualized. As urban scholarship expands, an understanding of the complexity of urban sustainability agendas emerged, shaped by both macro and micro-level ideologies and issues relating to cities' social, cultural, historical, ecological, material, and infrastructural functionality. A key literature-driven insight applicable to urban sustainability is urban resilience, and that we must attend to diverse experiences of cities and urban regions, understanding how these experiences differ significantly between spatial contexts and geographical settings. As an example of this diversity of experience with urban resilience, we analyzed the case of DMA within the urban nexus framework. The focus on DMA is particularly critical as it provides an avenue to not only understand urban resilience characterization in the GCC cities, but further portrays the complex interactions of coastal geography, industrial economy, unrelenting urban sprawl and oil sector led spatial planning that present both opportunities and challenges for urban resilience via nexus planning.

Our findings indicate that DMA is confronted with critical urbanization challenges producing significant environmental problems. The expansion of the petroleum industry in Saudi Arabia is intimately linked to the rise of DMA, producing employment opportunities and associated infrastructure and services (Abou-Korin & Al-Shihri, 2015). Yet, our analysis shows that the significant urban expansion in the city creating new industrial and commercial hubs is also producing urban sprawl negatively impacting on basic amenities and sanitation service provision. More worryingly, our findings show that DMA's urban expansion is so rapid that it surpasses the rate of population growth, significantly contributing to low-density, spreading urbanization (see also, Alhowaish, 2015). This situation raises considerable concerns for urban resilience as the urban environment is seriously threatened and the land for future development swallowed by uncontrolled urban expansion. The remaining acreage is insufficient to accommodate projected development demands as the ongoing subdivisions provide significant development obstacles to urban resilience efforts in DMA (Al-Shihri, 2016). Compounding land-use inadequacies are issues related to sustainable energy, safeguarded climate, secure food and sanitary water, and waste conditions (Hatab et al., 2019). How then do we hold together the urban development narrative in DMA to produce a resilient future? We discuss some urban resilience scholarly insights via the urban nexus framework.

As the DMA example demonstrates, the impacts of climate change and rapid urbanization on GCC cities are not even and involve significant socio-political and spatial contestation. Debates around the urban resilience question frequently investigate empirical conditions of urban transformation, but they did so specifically to understand the political stakes of urban change and general urban planning response. They were interested in what would happen to urban resilience when a focus on urban nexus planning framework is promoted and widely embraced. As a result, methods and procedures for addressing the implications of climate change and rapid urbanization have not adequately embraced nexus planning, such as autonomous techniques, proactive and reactive measures, long-term adaptation strategies, and transformational approaches (Darjee et al., 2023). Could nexus planning framework be a key partner in the pursuit of urban resilience, or would it turn out to be a political tool without delivering resilient benefits? This question, and its urban resilience implications, is just as significant across the global South, as it is in DMA as the race to address climate change and rapid urbanization problems become a defining urban transformation struggles of our times.

We have analyzed in this chapter a series of urban development conditions contributing to climate change, ranging from spatial expansion to natural ecosystem disturbance. DMA authorities are resisting and addressing these conditions through urban planning interventions. For example, the DMA has featured three development corridors or centers for growth, two of which are located along the two major roads that pass through it (UN-Habitat, 2019):

- The most significant highway in Saudi Arabia, Highway 40, which connects the major cities by traversing the whole nation and the Arabian Peninsula, has been designated as an institutional corridor. The 1359-km highway passes through Riyadh and links the West Coast cities of Jeddah and Makkah with the East Coast city of Dammam.
- The second corridor, which starts at the King Fahd Causeway, travels along Highway 95, often referred to as the Abu Haidriyah Highway. Major metropolitan areas like Jubail and DMA are accessible by the highway, which connects Bahrain with Saudi Arabia and stretches to the Kuwaiti border. The more than 300-km highway serves as the primary route for the transportation of vehicles and commodities in the region.
- The third axis runs along Highway 75, which links Al Hofuf and Dammam. The railway that links Riyadh with Dammam Port, Dammam City Center, and Al Hofuf runs beside this roadway. The main date producer in the area, AlAhsa Oasis, has farms and agricultural areas where this railway branches out.

Now and in the future, these urban development conditions do and will take multiple complex forms, which could incline toward weaking these planning interventions. For example, despite some urban planning efforts across Saudi Arabian cities the problems of rapid urbanization and climate change impacts remain dominant (Almulhim & Cobbinah, 2023, 2024). By recognizing the potential of nexus planning framework inherent in cities struggles against climate change and rapid urbanization, the framework articulates pathways toward managing urban development and climate change embedded within resilient and inclusive cities (Lehmann, 2018). However, as the case of DMA has highlighted, it is also often the case that nexus planning framework is not considered or wholly embraced by cities and hence contributes to reactive adaptation which threatens the pursuit of urban resilience (Carman & Zint, 2020; Teebken, 2018; Zhou et al., 2018).

The nexus planning framework offers extensive scenario planning strategy centered around goal-specified preliminary operations for adapting to climate change (Darjee et al., 2023) or managing rapid urbanization. The framework is primarily grounded in scientific data, with an emphasis on longer-term security of livelihood and functionality of cities. It is worth acknowledging, however, that the implementation of nexus planning framework calls for a sizable resource base, advanced technical expertise, and institutional ability, to deliberately guide choices about investments and policies that improve the flexibility for the adaptation of cities or any other arrangement involving the construction of small cities farther from city centers to lessen traffic and the strain on infrastructure. Founded on proactive initiatives, the nexus planning framework promotes sustainability and long-term resilience and are more effective in comparison to traditional planning. The framework is one of the transformative methods of urban adaptation that acknowledges the privileges, priorities, ideals, and weaknesses of previously excluded individuals and organizations. It prioritizes equal allocation of advantages and disadvantages of addressing climate change and managing rapid urbanization and takes marginalized groups into account during the planning and decision-making phases. Thus, in a rapidly changing urban landscape across the global South, the nexus planning framework offers transformative response to fast urbanization or climate change that entirely changes a system's basic features (Deubelli & Mechler, 2021; Panda, 2018).

Given this significance of the nexus planning framework to urban resilience, it is necessary to center it in investigating the possibilities of managing rapid urbanization and climate change management. Additionally, while the experiences of DMA with climate change impacts and urbanization challenges are quite complex, they are also mediated by existing planning and political context that have already shaped the urban sprawl and climate change experience across Saudi Arabia (Almulhim & Cobbinah, 2024). A nexus planning framework to urban transformation thus often responds to the ways that urban planning oriented toward inclusiveness and innovation intervene in these existing climate change-urbanization relations. Our analysis of DMA demonstrates that visions of urbanization and climate change management agendas in the GCC cities and across the global South that do not recognize these urbanization-climate change entanglements with long-standing implications on urban resilience risk reproducing and compounding these impacts and pressures.

8.6.2 Policy Prescriptions

Our findings highlighting rising land surface temperatures, increased air pollution, and ecological degradation caused by rapid urban expansion demonstrate the urgent need for proactive urban planning and governance policy response. We discuss here a number of policy prescriptions to guide DMA toward an inclusive and resilient future. First, it is recommended that DMA pursue decentralization of fiscal and jurisdictional powers, allowing the Amanah greater autonomy to implement nexus planning that focuses on localized, context-sensitive responses to social and spatial challenges, with a transfer of planning authority from MoMRA consistent with the principles of the New Urban Agenda. Through this process, the municipality should establish a systematic mechanism for collecting and analyzing de-identified demographic and environmental data to inform evidence-based adaptation strategies that address the impacts of climate-induced migration, long-term urban transformation, and improve nexus planning.

In a similar way, DMA should wholly adopt the urban nexus approach to support integrated and adaptive responses to climate change by identifying trade-offs and synergies across water, energy, and land systems, while aligning with scenario planning and circular economy strategies. This would promote the reuse and redevelopment of previously altered urban land, particularly within central zones, and to discourage unsustainable greenfield expansion as well as support spatial efficiency and long-term ecological preservation. With a focus on nexus planning, policymakers in DMA would naturally be encouraged to align urban transformation efforts with national frameworks such as Vision 2030, the National Spatial Strategy 2030, and SDG 11 to foster sustainable development, regional integration, and equitable access to safe, inclusive, and resilient urban environments.

Finally, DMA should institutionalize participatory planning mechanisms as part of efforts toward wholly embracing nexus planning to empower community stakeholders to actively engage in identifying priorities, shaping solutions, and evaluating outcomes. This has the potential to improving the responsiveness and effectiveness of urban policies. Coordination would further be strengthened particularly between metropolitan authorities through the development of reliable, cross-sector communication platforms. Through this process, knowledge sharing, improve implementation efficiency, and enhance collective governance capacity would be facilitated across the urban system.

8.7 Conclusion

Experiences described by the previous analyzed studies and supported by spatial data show that DMA is at a crossroads with significant urban resilience challenges. With rapid urbanization between 1993 and 2023, the metropolitan area was affected by considerable land-use challenges, pollution, and infrastructural shortfall. Urban planning, important for urban resilience and inclusive futures, was increasingly uncertain and increasingly inadequate due to the rapid expansion of DMA and severe impacts of climate change. Natural environment and ecosystems were also under threat due to their over-exploitation for urban development and changing weather patterns contributing to increased air pollutants in the metropolitan area. While the effects rapid urbanization, natural ecosystem conversion and increased pollution are becoming increasingly common in the DMA, our analysis shows that resilience efforts are hampered by limited engagement with the nexus planning approach. As a result, despite national and metropolitan planning efforts, these urban resilience challenges remain dominant. We developed a nexus framework to benefit transformative urban development strategies in DMA and beyond, as issues of climate change, urbanization, and urban resilience inadequacies are global, and particularly alarming in the global South.

As cities in the global South confront intensified and harsh realities of climate change and urbanization, urban resilience is a critical management approach to guide urban growth and expansion, address environmental degradation, and reduce pollution and related emissions. This capacity for urban resilience should be realized in a manner that addresses social inequalities and splintering urbanism and promotes an inclusive city environment. In this chapter, we have demonstrated that urban planners and urban studies scholars have a lot to contribute to this challenge, and that insights from the Middle East can help diverse understandings of urban resilience. Our analysis of DMA shows that we rethink urban resilience in ways that advance a more ecologically and socially inclusive city via urban nexus framework employing nexus planning, scenarios, modeling, and community engagement. We have discussed such an agenda for future research based on three adaptation pathways: reactive, planned, and transformative. Reactive adaptation points to response to urban problems particularly climate change when the impacts have occurred or occurring and fostering collective interests and collaboration. Planned adaptation draws on urban nexus framework to develop scenarios leading to proactive approach to managing urban growth and climate change threats. Transformational adaptation highlights the ways urban resilience can be applied, lived, encountered, and challenged.

Our argument for research diversification on urban resilience in the Middle East has the tendency to advance reinvigoration of knowledge production in urban planning in the region and global south more broadly. More importantly, it could lead to a closer collaboration between urban planners, local community, and scholars. City planners would benefit from better knowledge of and engagement with local community on the intricate but decisive details of urbanization, climate change impacts and threats, and governance inadequacies. Local communities would benefit from the critical policy initiatives focusing on building resilience via the urban nexus framework that extend beyond the city scale and networks. As our analysis illustrates, while the increasing policy attention on economic sustainability is acknowledged, it is worthwhile to demonstrate similar policy urgency on environmental sustainability in the short to long term through the urban nexus framework to guarantee a resilient future. Future research should incorporate institutional frameworks and stakeholder engagement, particularly with local governance bodies such as MoMRA, to complement and contextualize the spatial findings presented in this study. Future research should address the acknowledged limitation related to the centralized nature of urban governance in Saudi Arabia by incorporating institutional and policy analysis to enhance contextual understanding. Also, incorporate qualitative methods to capture institutional dynamics and stakeholder perspectives, thereby enriching the understanding of policy implementation and behavioral responses to urban resilience. In addition, comparative analyses with other Middle Eastern cities facing similar environmental and urbanization challenges would offer broader regional insights and enhance the generalizability of findings.

Acknowledgements This chapter is based on the following previously published article and is reproduced with permission from Elsevier: Almulhim, A. I., & Cobbinah, P. B. (2025). Dammam metropolitan area: Advancing the pursuit of urban resilience. Habitat International, 163, 103499.

References

Abdelrahman, K., Al-Otaibi, N., & Ibrahim, E. (2021). Assessment of land subsidence as an environmental threat facing Dammam city, eastern Saudi Arabia based on soil geotechnical parameters using downhole seismic approach. *Journal of King Saud University-Science, 33*(1), Article 101233.

AbouKorin, A. A., Alsayel, A., & Abdelfattah, H. M. (2020). Metropolitan Dammam: City of mega-projects. In S. Mehrotra, L. L. Lewis, M. Orloff, & B. Olberding (Eds.), *Volume II of greater than parts: A metropolitan opportunity. Washington, DC: World Bank. The Synthesis Report offers a range of integrated solutions (Mehrotra 2020)*, 5.

Abou-Korin, A. A., & Al-Shihri, F. S. (2015). Rapid urbanization and sustainability in Saudi Arabia: The case of Dammam metropolitan area. *Journal of Sustainable Development, 8*(9), 52. https://doi.org/10.5539/jsd.v8n9p52

Addas, A. (2023). Understanding the relationship between urban biophysical composition and land surface temperature in a hot desert megacity (Saudi Arabia). *International Journal of Environmental Research and Public Health, 20*(6), 5025.

Ahmed, M., Aloshan, M. A., Mohammed, W., Mesbah, E., Alsaleh, N. A., & Elghonaimy, I. (2024). Characterizing land surface temperature (LST) through remote sensing data for small-scale urban development projects in the gulf cooperation council (GCC). *Sustainability, 16*(9), 3873. https://doi.org/10.3390/su16093873

Aina, Y. A., Merwe, J. H. V., Alshuwaikhat, H. M. (2008, September 29–October 4). Urban spatial growth and land use change in Riyadh: Comparing spectral angle mapping and band ratioing techniques. In *Proceedings of the Academic Track of the 2008 Free and Open Source Software for Geospatial (FOSS4G) Conference, Incorporating the GISSA 2008 Conference, Cape Town, South Africa* (pp. 51–57).

Alatawi, A. S. (2022). Conservation action in Saudi Arabia: Challenges and opportunities. *Saudi Journal of Biological Sciences, 29*(5), 3466–3472.

Aldegheishem, A. (2024). Factors affecting ecological footprint in Saudi Arabia: A panel data analysis. *Frontiers in Environmental Science, 12*, 1384451.

Aldhafeeri, Z. M., & Alhazmi, H. (2022). Sustainability assessment of municipal solid waste in Riyadh, Saudi Arabia, in the framework of circular economy transition. *Sustainability, 14*(9), 5093. https://doi.org/10.3390/su14095093

Alghais, N., & Pullar, D. (2015). *Modelling the impacts of current patterns of urban form expansion in Kuwait with the use of ABM and GIS.* https://doi.org/10.36334/MODSIM.2015.L17.nasrin

Alghamdi, M. A. (2016). Characteristics and risk assessment of heavy metals in airborne PM 10 from a Residential Area of Northern Jeddah City, Saudi Arabia. *Polish Journal of Environmental Studies, 25*(3).

Alharbi, B., Shareef, M. M., & Husain, T. (2015). Study of chemical characteristics of particulate matter concentrations in Riyadh, Saudi Arabia. *Atmospheric Pollution Research, 6*(1), 88–98.

Alhowaish, A. K. (2015). Eighty years of urban growth and socioeconomic trends in Dammam Metropolitan Area, Saudi Arabia. *Habitat International, 50*, 90–98. https://doi.org/10.1016/j.habitatint.2015.08.019

Almheiri, A., Montenegro, J. F., Ewane, B. E., & Mohan, M. (2024). Climate change hazards and the resilience of coastal cities in the Gulf Cooperation Council countries: A systematic review. *City and Environment Interactions.* https://doi.org/10.1016/j.cacint.2024.100177

Almulhim, A. I. (2025). Building urban resilience through smart city planning: A systematic literature review. *Smart Cities, 8*(1), 22. https://doi.org/10.3390/smartcities8010022

Almulhim, A. I., & Al-Saidi, M. (2023). Circular economy and the resource nexus: Realignment and progress towards sustainable development in Saudi Arabia. *Environmental Development, 46*, Article 100851. https://doi.org/10.1016/j.envdev.2023.100851

Almulhim, A. I., & Cobbinah, P. B. (2023). Can rapid urbanization be sustainable? The case of Saudi Arabian cities. *Habitat International, 139*, Article 102884. https://doi.org/10.1016/j.habitatint.2023.102884

Almulhim, A. I., & Cobbinah, P. B. (2024). Framing resilience in Saudi Arabian cities: On climate change and urban policy. *Sustainable Cities and Society, 101*, Article 105172. https://doi.org/10.1016/j.scs.2024.105172

Almulhim, A. I., & Sharifi, A. (2025). Sustainable urban planning for addressing the compounded challenges of rapid urbanization and climate change. *Land Use Policy, 153*, Article 107544. https://doi.org/10.1016/j.landusepol.2025.107544

Alqurashi, A. F., & Kumar, L. (2014). Land use and land cover change detection in the Saudi Arabian desert cities of Makkah and Al-Taif using satellite data. *Advances in Remote Sensing, 3*(03), 106.

Alraouf, A. A. (2021). Towards a holistic interpretation of resilient cities' concept: The case of Doha, Qatar. *Journal of Design for Resilience in Architecture and Planning*. https://doi.org/10.47818/DRArch.2021.v2i2016

Alsharif, M., Alzandi, A. A., Shrahily, R., & Mobarak, B. (2022). Land use land cover change analysis for urban growth prediction using landsat satellite data and Markov Chain Model for Al Baha region Saudi Arabia. *Forests, 13*(10), 1530.

Alshehri, F., Abuamarah, B. A., & Abd El-Hamid, H. T. (2023). Impact of land use dynamics on land surface temperature using optical remote sensing data integrated with statistical analysis in Riyadh. *Saudi Arabia. Advances in Space Research, 72*(5), 1739–1750.

Al-Shihri, F. S. (2016). Impacts of large-scale residential projects on urban sustainability in Dammam Metropolitan Area, Saudi Arabia. *Habitat International, 56*, 201–211. https://doi.org/10.1016/j.habitatint.2016.06.001

Alwadei, M. M. (2022). *Air pollution climatology of particulate matter in Dammam, Saudi Arabia: Composition, sources and toxicity* (Doctoral dissertation, University of Birmingham).

Alwadei, M., Srivastava, D., Alam, M. S., Shi, Z., & Bloss, W. J. (2022). Chemical characteristics and source apportionment of particulate matter (PM2. 5) in Dammam, Saudi Arabia: Impact of dust storms. *Atmospheric Environment: X, 14*, 100164. https://doi.org/10.1016/j.aeaoa.2022.100164

Amenu, B. T., & Mamo, G. S. (2018). Review on wetland ecosystem destruction. *International Journal of Scientific Research in Civil Engineering, 2*(2), 5–15.

Amer, A. A. A. E., Meyad, A. M. E., Gao, Y., Niu, X., Chen, N., Xu, H., et al. (2022). Exploring the link between natural resources, urbanization, human capital, and ecological footprint: A case of GCC countries. *Ecological Indicators, 144*, Article 109556.

Amini, S., Saber, M., Rabiei-Dastjerdi, H., & Homayouni, S. (2022). Urban land use and land cover change analysis using random forest classification of landsat time series. *Remote Sensing, 14*(11), 2654.

Andersen, B., & Røe, P. G. (2016). The social context and politics of large scale urban architecture: Investigating the design of Barcode, Oslo. *European Urban and Regional Studies, 24*(3), 304–317. https://doi.org/10.1177/0969776416643751

Anguelovski, I., Connolly, J. J., Garcia-Lamarca, M., Cole, H., & Pearsall, H. (2018). New Scholarly Pathways on green gentrification: What does the urban 'green turn' mean and where is it going? *Progress in Human Geography, 43*(6), 1064–1086. https://doi.org/10.1177/0309132518803799

Aquilina, M. C., & Sheate, W. R. (2022). A critical analysis of the role of the urban climate resilience nexus in London. *European Planning Studies, 30*(7), 1355–1377. https://doi.org/10.1080/09654313.2021.1958758

Argyroudis, S. A., Mitoulis, S. A., Chatzi, E., Baker, J. W., Brilakis, I., Gkoumas, K., Vousdoukas, M., Hynes, W., Carluccio, S., Keou, O., Frangopol, D. M., & Linkov, I. (2022). Digital technologies can enhance climate resilience of critical infrastructure. *Climate Risk Management, 35*, 100387. https://doi.org/10.1016/j.crm.2021.100387

Artioli, F., Acuto, M., & McArthur, J. (2017). The water-energy-food nexus: An integration agenda and implications for urban governance. *Political Geography, 61*, 215–223. https://doi.org/10.1016/j.polgeo.2017.08.009

Breheny, M. J. (1992). *Sustainable development and urban form*. Pion.

Burton, E. (2000). The compact city: Just or just compact? A Preliminary Analysis. *Urban Studies, 37*(11), 1969–2006. https://doi.org/10.1080/00420980050162184

Canyon, D. (2021). Simplifying complexity with strategic foresight and scenario planning. *Security Nexus, 22*.

Carman, J. P., & Zint, M. T. (2020). Defining and classifying personal and household climate change adaptation behaviors. *Global Environmental Change, 61*, Article 102062. https://doi.org/10.1016/j.gloenvcha.2020.102062

Cervantes Puma, G. C., Salles, A., & Bragança, L. (2024). Nexus between urban circular economies and sustainable development goals: A systematic literature review. *Sustainability, 16*(6), 2500. https://doi.org/10.3390/su16062500

Chen, B., & Lu, Y. (2015). Urban nexus: A new paradigm for urban studies. *Ecological Modelling, 318*, 5–7. https://doi.org/10.1016/j.ecolmodel.2015.10.010

Cobbinah, P. B. (2021). Urban resilience in climate change hotspot. *Land Use Policy, 100*, Article 104948.

Crippa, M., Guizzardi, D., Pagani, F., Banja, M., Muntean, M., Schaaf, E., et al. (2023). *GHG emissions of all world countries*. Publications Office of the European Union.

Darjee, K. B., Neupane, P. R., & Köhl, M. (2023). Proactive adaptation responses by vulnerable communities to climate change impacts. *Sustainability, 15*(14), 10952. https://doi.org/10.3390/su151410952

De Jesus, J. B., & Santana, I. D. (2017). Estimation of land surface temperature in caatinga area using Landsat 8 data. *Journal of Hyperspectral Remote Sensing, 7*(3), 150–157.

Deng, X., Zhao, C., & Yan, H. (2013). Systematic modeling of impacts of land use and land cover changes on regional climate: A review. *Advances in Meteorology, 2013*(1), Article 317678.

Deubelli, T. M., & Mechler, R. (2021). Perspectives on transformational change in climate risk management and adaptation. *Environmental Research Letters, 16*(5), Article 053002. https://doi.org/10.1088/1748-9326/abd42d

Emamjomehzadeh, O., Kerachian, R., Emami-Skardi, M. J., & Momeni, M. (2023). Combining urban metabolism and reinforcement learning concepts for sustainable water resources management: A nexus approach. *Journal of Environmental Management, 329*, Article 117046. https://doi.org/10.1016/j.jenvman.2022.117046

Farahat, A. (2016). Air pollution in the Arabian Peninsula: Causes, effects, and aerosol categorization. *Arabian Journal of Geosciences, 9*, 1–17. https://doi.org/10.1007/s12517-015-2203-y

Fattah, M. A., Morshed, S. R., & Morshed, S. Y. (2021). Impacts of land use-based carbon emission pattern on surface temperature dynamics: Experience from the urban and suburban areas of Khulna, Bangladesh. *Remote Sensing Applications: Society and Environment, 22*, Article 100508. https://doi.org/10.1016/j.rsase.2021.100508

Fluet-Chouinard, E., Stocker, B. D., Zhang, Z., Malhotra, A., Melton, J. R., Poulter, B., Kaplan, J. O., Goldewijk, K. K., Siebert, S., Minayeva, T., Hugelius, G., & McIntyre, P. B. (2023). Extensive global wetland loss over the past three centuries. *Nature, 614*(7947), 281–286.

Gascon, M., Cirach, M., Martínez, D., Dadvand, P., Valentín, A., Plasència, A., & Nieuwenhuijsen, M. J. (2016). Normalized difference vegetation index (NDVI) as a marker of surrounding greenness in epidemiological studies: The case of Barcelona city. *Urban Forestry & Urban Greening, 19*, 88–94. https://doi.org/10.1016/j.ufug.2016.07.001

Ghisellini, P., Cialani, C., & Ulgiati, S. (2016). A review on circular economy: The expected transition to a balanced interplay of environmental and economic systems. *Journal of Cleaner Production, 114*, 11–32. https://doi.org/10.1016/j.jclepro.2015.09.007

Giatti, L. L., Lazaro, L. L. B., Amaral, M. H., dos Santos, K. L., Matos, A. C., de Lima, V. R. S., & Urbinatti, A. M. (2024). Nexus for urban resilience in the face of climate change: Policies and synergies in the context of a macrometropolis. In *Environmental governance of the São Paulo macrometropolis: Perspectives on climate variability* (pp. 21–34). Springer Nature Switzerland. https://doi.org/10.1007/978-3-031-59611-7_2

Gondhalekar, D., & Erlbeck, R. (2021). Application of the nexus approach as an integrated urban planning framework: From theory to practice and back again. *Frontiers in Sustainable Cities, 3*, Article 751682. https://doi.org/10.3389/frsc.2021.751682

Haddad, N. M., Brudvig, L. A., Clobert, J., Davies, K. F., Gonzalez, A., Holt, R. D., Lovejoy, T. E., Sexton, J. O., Austin, M. P., Collins, C. D., Cook, W. M., & Townshend, J. R. (2015). Habitat fragmentation and its lasting impact on Earth's ecosystems. *Science Advances, 1*(2), e1500052. https://doi.org/10.1126/sciadv.1500052

Harrison, R. M., Bousiotis, D., Mohorjy, A. M., Alkhalaf, A. K., Shamy, M., Alghamdi, M., Khoder, M., & Costa, M. (2017). Health risk associated with airborne particulate matter and its components in Jeddah, Saudi Arabia. *Science of the Total Environment, 590*, 531–539.

Hasan, S. S., Zhen, L., Miah, M. G., Ahamed, T., & Samie, A. (2020). Impact of land use change on ecosystem services: A review. *Environmental Development, 34*, Article 100527.

Hassan, R., Rahman, M., & Hamdan, A. (2022, May). Assessment of air quality index (AQI) in Riyadh, Saudi Arabia. In *IOP Conference Series: Earth and Environmental Science* (Vol. 1026, No. 1, p. 012003). IOP Publishing.

Hatab, A. A., Cavinato, M. E. R., Lindemer, A., & Lagerkvist, C. J. (2019). Urban sprawl, food security and agricultural systems in developing countries: A systematic review of the literature. *Cities, 94*, 129–142. https://doi.org/10.1016/j.cities.2019.06.001

Hettiarachchi, H., & Ardakanian, R. (2016). *Managing water, soil, and waste in the context of global change* (pp. 1–7). Springer International Publishing. https://doi.org/10.1007/978-3-319-28593-1_1

Heydkamp, C., Braun, S., & Radecki, A. V. (2014). Morgenstadt benchmarking and city development concepts. Integrated resource management in Asian cities: The urban nexus and sustainable urban governance. https://hdl.handle.net/20.500.12870/3792

Hoekstra, A. Y., Buurman, J., & Van Ginkel, K. C. (2018). Urban water security: A review. *Environmental Research Letters, 13*(5), Article 053002. https://doi.org/10.1088/1748-9326/aaba52

Huang, S., Suo, C., Guo, J., Lv, J., Jing, R., Yu, L., Fan, Y., & Ding, Y. (2023). Balancing the water-energy dilemma in nexus system planning with bi-level and multi-uncertainty. *Energy*, 127720. https://doi.org/10.1016/j.energy.2023.127720

Hulley, G. C., Ghent, D., Göttsche, F. M., Guillevic, P. C., Mildrexler, D. J., & Coll, C. (2019). Land surface temperature. In *Taking the temperature of the earth* (pp. 57–127). Elsevier.

Hussain, S., Mubeen, M., Ahmad, A., Majeed, H., Qaisrani, S. A., Hammad, H. M., Amjad, M., Ahmad, I., Fahad, S., Ahmad, N., & Nasim, W. (2023). Assessment of land use/land cover changes and its effect on land surface temperature using remote sensing techniques in Southern Punjab, Pakistan. *Environmental Science and Pollution Research, 30*(44), 99202–99218.

Hussein, T., Alghamdi, M. A., Khoder, M., AbdelMaksoud, A. S., Al-Jeelani, H., Goknil, M. K., Shabbaj, I. I., Almehmadi, F. M., Hyvärinen, A., Lihavainen, H., & Hämeri, K. (2014). Particulate matter and number concentrations of particles larger than 0.25 μm in the urban atmosphere of Jeddah, Saudi Arabia. *Aerosol and Air Quality Research, 14*(5), 1383–1391.

Javan, K., Altaee, A., BaniHashemi, S., Darestani, M., Zhou, J., & Pignatta, G. (2023). A review of interconnected challenges in the water–energy–food nexus: Urban pollution perspective towards sustainable development. *Science of the Total Environment*, 169319. https://doi.org/10.1016/j.scitotenv.2023.169319

Jenks, M., Burton, E., & Williams, K. (Eds.). (1996). *The compact city: A sustainable urban form?* E. & F.N. Spon. London.

Jennings, V., Browning, M. H., Rigolon, A., Jennings, V., Browning, M. H., & Rigolon, A. (2019). Urban green space at the nexus of environmental justice and health equity. In *Urban green spaces: Public health and sustainability in the United States* (pp. 47–69). https://doi.org/10.1007/978-3-030-10469-6_4

Jimenez-Munoz, J. C., Sobrino, J. A., Skoković, D., Mattar, C., & Cristobal, J. (2014). Land surface temperature retrieval methods from Landsat-8 thermal infrared sensor data. *IEEE Geoscience and Remote Sensing Letters, 11*(10), 1840–1843.

Johnson, O. W., & Karlberg, L. (2017). Co-exploring the water-energy-food nexus: Facilitating dialogue through participatory scenario building. *Frontiers in Environmental Science, 5*, 24. https://doi.org/10.3389/fenvs.2017.00024

Kim, H., Yoon, S., Cho, E., & Kim, J. (2015). *A study on policy directions for the water-food-energy nexus (I) water-energy nexus*. Network of Institutions for Sustainable Development.

Kolo, J., Schrouder, S., & Almashhour, R. (2021). Deconstructing the transport-health nexus: Urban planning roles and dividends for cities. *Journal of Transport & Health, 22*, Article 101202. https://doi.org/10.1016/j.jth.2021.101202

Koutra, S., Mondejar, M. B., & Becue, V. (2022). The nexus of 'urban resilience'and 'energy efficiency'in cities. *Current Research in Environmental Sustainability, 4*, Article 100118. https://doi.org/10.1016/j.crsust.2021.100118

Labban, A. H., & Butt, M. J. (2023). Evaluation of MERRA-2 data for aerosols patterns over the Kingdom of Saudi Arabia. *Heliyon, 9*(6).

Lehmann, S. (2018). Implementing the urban nexus approach for improved resource-efficiency of developing cities in Southeast-Asia. *City, Culture and Society, 13*, 46–56. https://doi.org/10.1016/j.ccs.2017.10.003

Li, G., Fang, C., & Qi, W. (2021). Different effects of human settlements changes on landscape fragmentation in China: Evidence from grid cell. *Ecological Indicators, 129*, Article 107927. https://doi.org/10.1016/j.ecolind.2021.107927

Li, X., Stringer, L. C., & Dallimer, M. (2022). The impacts of urbanisation and climate change on the urban thermal environment in Africa. *Climate, 10*(11), 164. https://doi.org/10.3390/cli10110164

Li, Z. L., Tang, B. H., Wu, H., Ren, H., Yan, G., Wan, Z., Trigo, I. F., & Sobrino, J. A. (2013). Satellite-derived land surface temperature: Current status and perspectives. *Remote Sensing of Environment, 131*, 14–37.

Liang, M. S., Huang, G. H., Chen, J. P., & Li, Y. P. (2022). Development of non-deterministic energy-water-carbon nexus planning model: A case study of Shanghai, China. *Energy, 246*, Article 123300. https://doi.org/10.1016/j.energy.2022.123300

Lim, C. C., Thurston, G. D., Shamy, M., Alghamdi, M., Khoder, M., Mohorjy, A. M., Alkhalaf, A. K., Brocato, J., Chen, L. C., & Costa, M. (2018). Temporal variations of fine and coarse particulate matter sources in Jeddah, Saudi Arabia. *Journal of the Air & Waste Management Association, 68*(2), 123–138.

Mahmood, H., Alkhateeb, T. T. Y., Al-Qahtani, M. M. Z., Allam, Z., Ahmad, N., & Furqan, M. (2019). Agriculture development and CO2 emissions nexus in Saudi Arabia. *PLoS One, 14*(12), Article e0225865. https://doi.org/10.1371/journal.pone.0225865

Mallick, J., Singh, V. P., Almesfer, M. K., Talukdar, S., Alsubhi, M., Ahmed, M., & Khan, R. A. (2022). Spatio-temporal analysis and simulation of land cover changes and their impacts on land surface temperature in urban agglomeration of Bisha Watershed, Saudi Arabia. *Geocarto International, 37*(25), 7591–7617.

Meerow, S., Newell, J. P., & Stults, M. (2016). Defining urban resilience: A review. *Landscape and Urban Planning, 147*, 38–49.

Mileti, D. (1999). *Disasters by design*. Joseph Henry Press.

Mohammad, A., Alzandi, A. A., Shrahily, R., & Mobarak, B. (2022). Land use land cover change analysis for urban growth prediction using landsat satellite data and Markov chain model for Al Baha region Saudi Arabia. *Forests, 2022*(13), 1530.

Monstadt, J., & Coutard, O. (2019). Cities in an era of interfacing infrastructures: Politics and spatialities of the urban nexus. *Urban Studies, 56*(11), 2191–2206. https://doi.org/10.1177/0042098019833907

Næss, P. (2006). *Urban structure matters: Residential location, car dependence and travel behaviour*. Routledge.

Nagendra, H., Bai, X., Brondizio, E. S., & Lwasa, S. (2018). The urban south and the predicament of global sustainability. *Nature Sustainability, 1*(7), 341–349. https://doi.org/10.1038/s41893-018-0101-5

Nguyen, H. T. T., Doan, T. M., & Radeloff, V. (2018). Applying random forest classification to map land use/land cover using Landsat 8 OLI. *The International Archives of the Photogrammetry, Remote Sensing and Spatial Information Sciences, 42*, 363–367.

Nhamo, L., Mabhaudhi, T., Mpandeli, S., Dickens, C., Nhemachena, C., Senzanje, A., Naidoo, D., Liphadzi, S., & Modi, A. T. (2020). An integrative analytical model for the water-energy-food nexus: South Africa case study. *Environmental Science & Policy, 109*, 15–24. https://doi.org/10.1016/j.envsci.2020.04.010

Nhamo, L., Mpandeli, S., & Mabhaudhi, T. (2023). Achieving urban resilience and sustainability through nexus planning. *SABI Magazine-Tydskrif, 16*(1), 32–33. https://hdl.handle.net/10520/ejc-sa_bimag_v16_n1_a19

Nhamo, L., & Ndlela, B. (2021). Nexus planning as a pathway towards sustainable environmental and human health post Covid-19. *Environmental Research, 192*, Article 110376. https://doi.org/10.1016/j.envres.2020.110376

O'Neill, B. C., Kriegler, E., Riahi, K., Ebi, K. L., Hallegatte, S., Carter, T. R., Mathur, R., & van Vuuren, D. P. (2014). A new scenario framework for climate change research: The concept of shared socioeconomic pathways. *Climatic Change, 122*(3), 387–400. https://doi.org/10.1007/s10584-013-0905-2

Opiyo, P. O., & Agong, S. G. (2020). Nexus between urban food system and other urban systems: Exploring opportunities for improving food security in Kisumu. *Kenya*. https://doi.org/10.12691/seg-5-1-4

Panda, A. (2018). Transformational adaptation of agricultural systems to climate change. *Wiley Interdisciplinary Reviews: Climate Change, 9*(4), Article e520. https://doi.org/10.1002/wcc.520

Porter, W. C., Khalil, M. A. K., Butenhoff, C. L., Almazroui, M., Al-Khalaf, A. K., & Al-Sahafi, M. S. (2014). Annual and weekly patterns of ozone and particulate matter in Jeddah, Saudi Arabia. *Journal of the Air & Waste Management Association, 64*(7), 817–826.

Raggad, B. (2018). Carbon dioxide emissions, economic growth, energy use, and urbanization in Saudi Arabia: Evidence from the ARDL approach and impulse saturation break tests. *Environmental Science and Pollution Research, 25*, 14882–14898.

Rahman, M. T., Aldosary, A. S., & Mortoja, M. G. (2017). Modeling future land cover changes and their effects on the land surface temperatures in the Saudi Arabian eastern coastal city of Dammam. *Land, 6*(2), 36. https://doi.org/10.3390/land6020036

Raihan, A., & Tuspekova, A. (2022). Dynamic impacts of economic growth, energy use, urbanization, agricultural productivity, and forested area on carbon emissions: New insights from Kazakhstan. *World Development Sustainability, 1*, Article 100019. https://doi.org/10.1016/j.wds.2022.100019

Rajeshwari, A., & Mani, N. D. (2014). Estimation of land surface temperature of Dindigul district using Landsat 8 data. *International Journal of Research in Engineering and Technology, 3*(5), 122–126.

Rehan, M., & Munir, S. (2022). Analysis and modeling of air pollution in extreme meteorological conditions: A case study of Jeddah, the Kingdom of Saudi Arabia. *Toxics, 10*(7), 376. https://doi.org/10.3390/toxics10070376

Rodriguez-Galiano, V. F., Ghimire, B., Rogan, J., Chica-Olmo, M., & Rigol-Sanchez, J. P. (2012). An assessment of the effectiveness of a random forest classifier for land-cover classification. *ISPRS Journal of Photogrammetry and Remote Sensing, 67*, 93–104.

Safieddine, S., Clerbaux, C., Clarisse, L., Whitburn, S., & Eltahir, E. A. (2022). Present and future land surface and wet bulb temperatures in the Arabian Peninsula. *Environmental Research Letters, 17*(4), Article 044029.

Sager, T. (2014). Ideological traces in plans for compact cities: Is neo-liberalism hegemonic? *Planning Theory, 14*(3), 268–295. https://doi.org/10.1177/1473095214527279

Sarikhani, A., Joda, F., Ledari, M. B., & Akbarnavasi, H. (2023). Fostering self-sustaining water-energy regions: A nexus approach to mitigate water scarcity. *Sustainable Production and Consumption, 43*, 230–250. https://doi.org/10.1016/j.spc.2023.10.021

Schmidt, M. (2021). The resource-energy nexus as a key factor for circular economy. *Chemie Ingenieur Technik, 93*(11), 1707–1716. https://doi.org/10.1002/cite.202100111

Sekertekin, A., & Zadbagher, E. (2021). Simulation of future land surface temperature distribution and evaluating surface urban heat island based on impervious surface area. *Ecological Indicators, 122*, Article 107230.

Shih, H. C., Stow, D. A., Chang, K. C., Roberts, D. A., & Goulias, K. G. (2022). From land cover to land use: Applying random forest classifier to Landsat imagery for urban land-use change mapping. *Geocarto International, 37*(19), 5523–5546.

Shirazi, S. A., & Kazmi, J. H. (2016). Analysis of socio-environmental impacts of the loss of urban trees and vegetation in Lahore, Pakistan: A review of public perception. *Ecological Processes, 5*, 1–12.

Sim, L.-C. (2023). Renewable energy and governance resilience in the Gulf. *Energies, 16*(7), Article 3225. https://doi.org/10.3390/en16073225

Sobrino, J. A., Jiménez-Muñoz, J. C., & Paolini, L. (2004). Land surface temperature retrieval from LANDSAT TM 5. *Remote Sensing of Environment, 90*(4), 434–440.

Song, X. P., Hansen, M. C., Stehman, S. V., Potapov, P. V., Tyukavina, A., Vermote, E. F., & Townshend, J. R. (2018). Global land change from 1982 to 2016. *Nature, 560*(7720), 639–643. https://doi.org/10.1038/s41586-018-0411-9

Sperling, J. B., & Berke, P. R. (2017). Urban nexus science for future cities: Focus on the energy-water-food-X nexus. *Current Sustainable/renewable Energy Reports, 4*, 173–179. https://doi.org/10.1007/s40518-017-0085-1

Sresto, M. A., Siddika, S., Fattah, M. A., Morshed, S. R., & Morshed, M. M. (2022). A GIS and remote sensing approach for measuring summer–winter variation of land use and land cover indices and surface temperature in Dhaka district, Bangladesh. *Heliyon, 8*(8), Article e10309. https://doi.org/10.1016/j.heliyon.2022.e10309

Stevens, M. (2021). *How does circularity bring resilience to cities? Understanding urban circularity*. http://hdl.handle.net/2268.2/13591.

Su, Q., Chen, K., & Liao, L. (2021). The impact of land use change on disaster risk from the perspective of efficiency. *Sustainability, 13*(6), 3151. https://doi.org/10.3390/su13063151

Teebken, J. (2018). *Ten years, nine floods: Local-level climate adaptation in China*. https://www.newsecuritybeat.org/2018/03/ten-years-floods-local-level-climate-adaptation-china/

Tikuye, B. G., Gill, L., Rusnak, M., & Manjunatha, B. R. (2023). Modelling the impacts of changing land use and climate on sediment and nutrient retention in Lake Tana Basin, Upper Blue Nile River Basin, Ethiopia. *Ecological Modelling, 482*, 110383. https://doi.org/10.1016/j.ecolmodel.2023.110383

Turnheim, B., Berkhout, F., Geels, F. W., Hof, A., McMeekin, A., Nykvist, B., & van Vuuren, D. P. (2015). Evaluating sustainability transitions pathways: Bridging analytical approaches to address governance challenges. *Global Environmental Change, 35*, 239–253. https://doi.org/10.1016/j.gloenvcha.2015.08.010

Uddin, M. S., Routray, J. K., & Warnitchai, P. (2019). Systems thinking approach for resilient critical infrastructures in urban disaster management and sustainable development. *Resilient Structures and Infrastructure*, 379–415. https://doi.org/10.1007/978-981-13-7446-3_15

UNDP. (2021). *National Spatial Strategy 2030*. United Nations Development Programme. https://www.undp.org/saudi-arabia/blog/national-spatial-strategy-2030

UN-ESCAP. (2016). *Nexus: Conceptual framework and linkages to global agendas*. https://www.unescap.org/sites/default/files/The%20Urban%20Nexus_First%20Draft.pdf

UN-Habitat. (2019). *DAMMAM city profile. United Nations Human Settlements Programme (UN-Habitat)*. https://unhabitat.org/sites/default/files/2020/04/dammam.pdf

Urbinatti, A. M., Benites-Lazaro, L. L., Carvalho, C. M. D., & Giatti, L. L. (2020). The conceptual basis of water-energy-food nexus governance: Systematic literature review using network and discourse analysis. *Journal of Integrative Environmental Sciences, 17*(2), 21–43. https://doi.org/10.1080/1943815X.2020.1749086

Van Soest, H. L., Van Vuuren, D. P., Hilaire, J., Minx, J. C., Harmsen, M. J., Krey, V., Popp, A., Riahi, K., & Luderer, G. (2019). Analysing interactions among Sustainable Development Goals with Integrated Assessment Models. *Global Transitions, 1*, 210–225. https://doi.org/10.1016/j.glt.2019.10.004

Vision 2030. (2018). *Kingdom of Saudi Arabia*. http://vision2030.gov.sa

Wigginton, N. S., Fahrenkamp-Uppenbrink, J., Wible, B., & Malakoff, D. (2016). Cities are the future. *Science, 352*(6288), 904–905. https://doi.org/10.1126/science.352.6288.904

Xue, J., Liu, G., Casazza, M., & Ulgiati, S. (2018). Development of an urban FEW nexus online analyzer to support urban circular economy strategy planning. *Energy, 164*, 475–495. https://doi.org/10.1016/j.energy.2018.08.198

Yang, M., Chen, L., Wang, J., Msigwa, G., Osman, A. I., Fawzy, S., Rooney, D. W., & Yap, P. S. (2023). Circular economy strategies for combating climate change and other environmental issues. *Environmental Chemistry Letters, 21*(1), 55–80. https://doi.org/10.1007/s10311-022-01499-6

Yengoh, G. T., Dent, D., Olsson, L., Tengberg, A. E., & Tucker, C. J., III. (2015). Use of the normalized difference vegetation index (NDVI) to assess land degradation at multiple scales: Current status, future trends, and practical considerations. *SpringerBriefs in Environmental Science*. https://doi.org/10.1007/978-3-319-24112-8

Zhou, Q., Leng, G., & Huang, M. (2018). Impacts of future climate change on urban flood volumes in Hohhot in northern China: Benefits of climate change mitigation and adaptations. *Hydrology and Earth System Sciences, 22*(1), 305–316. https://doi.org/10.5194/hess-22-305-2018

Chapter 9
Planning for Sustainable Urban Futures: Rethinking Urbanization and Resilience in Saudi Arabia

9.1 Urbanization under Arid Constraint: Core Insights

This book has demonstrated that urbanization in Saudi Arabia must be understood as a process unfolding under persistent and non-negotiable environmental constraints. Unlike urban contexts where climate pressures are episodic or emerging, Saudi cities have developed within conditions of extreme heat, chronic water scarcity, and high resource intensity for decades. These conditions are not external shocks to which cities occasionally respond; they are structural parameters that shape everyday urban life and long-term development trajectories.

Across the chapters, a central insight emerges: rapid urbanization under arid conditions fundamentally alters the risk profile of cities. Patterns of spatial expansion, infrastructure provision, and service delivery that may appear manageable in resource-abundant environments generate cumulative vulnerabilities when environmental margins are narrow. Urban sprawl, for example, is not merely a land-use inefficiency in Saudi Arabia; it increases exposure to heat, raises infrastructure maintenance costs, and intensifies dependence on energy- and water-intensive systems. Similarly, infrastructure stress reflects not only population growth but also the environmental cost of sustaining urban systems under extreme climatic conditions.

The Saudi experience highlights the limits of growth-oriented urbanization models that prioritize speed, scale, and economic output without sufficiently integrating environmental realities into planning decisions. While economic development has enabled rapid improvements in housing, mobility, and services, it has also produced urban forms and infrastructure systems whose long-term sustainability is increasingly uncertain. This book argues that these outcomes are not the inevitable consequence of rapid growth, but the result of planning approaches that have historically treated environmental constraint as a secondary consideration rather than as a foundational planning parameter.

In this sense, Saudi urbanization serves as a revealing case for understanding how aridity reshapes the meaning of sustainability and resilience. It demonstrates that

A. I. Almulhim and P. B. Cobbinah, *Urbanization Rethought*, Urban Sustainability,
https://doi.org/10.1007/978-981-92-0118-1_9

environmental constraint is not simply an additional variable to be managed, but a condition that redefines the trade-offs inherent in urban development. Recognizing this reality is essential for any meaningful rethinking of urban futures in arid and climate-constrained contexts.

9.2 Planning as a Mediating Practice, Not a Technical Tool

A core argument advanced throughout this book is that urban planning functions as a mediating practice between development ambition and environmental constraint. Planning is not merely a technical exercise concerned with zoning, regulations, or spatial layouts; it is an institutional process through which competing priorities, economic growth, environmental protection, social equity, and political objectives are negotiated over time. In Saudi Arabia, the evolving role of planning reveals both the risks of its marginalization and the potential of its re-centering.

Historically, planning in Saudi cities often followed development rather than shaping it. Large-scale investments, sectoral priorities, and administrative decisions drove urban expansion, with planning frameworks adapting reactively to accommodate growth. This sequencing limited the ability of planning to guide urban form proactively, contributing to fragmented land-use patterns and infrastructure systems that are difficult to retrofit under arid conditions. The consequences of this approach have become increasingly apparent as environmental pressures intensify, and urban systems approach operational limits.

The book's analysis shows that planning is now being repositioned as a strategic instrument within Saudi urban governance. Climate risk, sustainability concerns, and the complexity of contemporary urban development have renewed attention to the role of planning in coordinating across sectors and scales. However, this repositioning remains uneven. Planning institutions are often expected to deliver integrated, resilience-oriented outcomes without corresponding expansion in authority, resources, or institutional coherence.

This tension underscores a critical insight: planning cannot mediate effectively between ambition and constraint if it is treated as a procedural requirement rather than as a governing practice. In arid environments, where the costs of misalignment are amplified, planning decisions related to density, infrastructure alignment, and land-use integration have long-term implications for resource consumption and environmental exposure. The Saudi experience demonstrates that strengthening planning capacity is not simply a matter of improving technical expertise, but of enhancing institutional coordination and clarifying governance responsibilities.

By conceptualizing planning as a mediating practice, this book shifts attention from individual projects or policies to the cumulative effects of planning decisions over time. It highlights how planning shapes pathways of urban development that can either entrench vulnerability or enable adaptive capacity. This perspective is central to rethinking urban resilience in Saudi Arabia and beyond.

9.3 Sustainability, Growth, and the Limits of Reactive Urbanism

The question of whether rapid urbanization can be sustainable has been a recurring theme throughout this book. Chapters examining patterns of urban growth in Saudi cities reveal that sustainability challenges are not simply the result of insufficient environmental awareness, but of development models that prioritize reactive responses over anticipatory planning. Urban expansion has often been addressed through incremental infrastructure provision and regulatory adjustments rather than through strategic reconsideration of urban form and growth patterns.

This reactive approach has produced a range of sustainability challenges. Infrastructure systems have been stretched to accommodate dispersed development, increasing operational costs and resource consumption. Socio-spatial inequalities have emerged as access to services and environmental quality varies across urban areas. Environmental degradation, manifested in heat accumulation, water stress, and ecological disruption, has compounded these challenges, particularly in cities experiencing rapid demographic change.

The Saudi case illustrates that sustainability cannot be retrofitted easily once urban form and infrastructure systems are established. In arid environments, where resource constraints are acute, the window for corrective intervention is often narrow. This reality challenges narratives that frame sustainability as a set of add-on measures or technological fixes. Instead, the book argues for a proactive planning orientation that integrates sustainability considerations into core decisions about land use, density, mobility, and infrastructure coordination from the outset.

Importantly, the book does not suggest that growth and sustainability are inherently incompatible. Rather, it demonstrates that their compatibility depends on how growth is planned and governed. Compact urban form, integrated transport systems, and inclusive planning mechanisms emerge not as universal solutions, but as context-sensitive strategies whose relevance and effectiveness depend on environmental and institutional conditions. The Saudi experience underscores the need to move beyond reactive urbanism toward planning approaches that anticipate environmental limits and manage growth accordingly.

9.4 Rethinking Urban Resilience in Climate-Constrained Cities

Urban resilience has become a prominent concept in contemporary urban discourse, particularly in the context of climate change. In Saudi Arabia, resilience has been increasingly invoked in relation to risks such as extreme heat, water scarcity, flooding, and dust storms. While this shift reflects growing recognition of climate vulnerability, the book has argued that resilience must be examined critically as a planning practice rather than accepted unproblematically as a policy goal.

A key insight emerging from the book is that resilience in arid environments is fundamentally different from resilience in contexts characterized by episodic shocks. In Saudi cities, environmental stress is continuous rather than exceptional. Heat, water scarcity, and resource dependence exert persistent pressure on urban systems, shaping everyday conditions rather than discrete crises. As a result, resilience cannot be reduced to the capacity to recover from isolated events; it must encompass the ability of urban systems to function and adapt under chronic stress.

The analysis shows that resilience strategies in Saudi Arabia often emphasize robustness and protection, focusing on infrastructure hardening or risk mitigation. While such measures are important, they may be insufficient if they do not address underlying exposure and vulnerability. Without careful integration into planning frameworks, resilience interventions risk reinforcing existing development patterns that generate long-term risk.

This book therefore advocates a shift toward resilience as adaptive urban transformation. Such an approach prioritizes flexibility, learning, and long-term stewardship over short-term performance. It recognizes uncertainty as an inherent condition of climate-constrained urbanization and emphasizes the role of planning institutions in managing this uncertainty. By situating resilience within the broader planning system, the book contributes to a more nuanced understanding of how cities in arid environments can navigate environmental constraint without relying on simplified or universal frameworks.

9.5 Green Infrastructure and Megaprojects as Conditional Planning Instruments

One of the central contributions of this book lies in its treatment of green infrastructure and megaprojects not as solutions in themselves, but as conditional planning instruments whose performance is shaped by environmental context, institutional capacity, and governance arrangements. In contemporary Saudi urbanization, both approaches have assumed prominent roles within planning and policy discourse. Yet their implications for urban resilience under arid conditions remain deeply contingent.

Green infrastructure has gained traction as a means of addressing climate stress, enhancing liveability, and supporting resilience. However, as demonstrated in Chap. 6, its applicability in arid environments cannot be assumed. In Saudi cities, green infrastructure operates within a tightly coupled water–energy–urban planning nexus, where irrigation demand, energy consumption, and long-term maintenance present significant challenges. The book's analysis reveals that green infrastructure contributes to resilience only when it is strategically planned, selectively deployed, and aligned with long-term resource availability. When implemented symbolically or as a replication of models developed in temperate contexts, it risks exacerbating resource dependency and operational fragility.

This conditionality is mirrored in the case of megaproject-led urban development. Megaprojects have emerged as defining features of Saudi urban transformation, offering opportunities for integrated planning, large-scale infrastructure coordination, and experimentation with new urban forms. At the same time, their scale magnifies the consequences of planning decisions, particularly under conditions of environmental constraint. As Chap. 7 has shown, megaprojects can embed rigid spatial and infrastructural configurations that limit adaptability and increase long-term exposure to climate risk.

The juxtaposition of green infrastructure and megaprojects underscores a broader insight: resilience is not inherent to specific planning tools. Rather, it emerges, or fails to emerge, from how such tools are embedded within planning systems, governed over time, and adjusted in response to environmental feedback. In arid contexts, where margins for error are narrow, the stakes of misalignment are especially high. This book therefore cautions against treating either green infrastructure or megaprojects as universally desirable interventions, emphasizing instead the need for planning-led evaluation grounded in environmental realism.

9.6 Governance, Institutions, and the Implementation Gap

Throughout the book, a recurring theme has been the gap between planning ambition and implementation capacity. This gap is particularly pronounced in arid environments, where the complexity of managing water, energy, land use, and climate risk demands high levels of institutional coordination. In Saudi Arabia, evolving governance structures have sought to address this complexity, yet the book's analysis reveals persistent challenges in translating strategic visions into operational resilience.

Planning and resilience agendas in Saudi cities often involve multiple actors operating across different scales, from national ministries to regional authorities and municipal agencies. While this multi-level governance structure allows for strategic alignment, it also introduces coordination challenges that can undermine implementation. Responsibilities for planning, infrastructure provision, environmental management, and service delivery are frequently fragmented, creating institutional silos that complicate integrated action.

The book demonstrates that governance capacity is not simply a function of institutional design, but of how roles, incentives, and accountability mechanisms are aligned over time. In the context of megaprojects, for example, governance arrangements may prioritize delivery speed and investment attraction, potentially sidelining long-term operational considerations. Similarly, green infrastructure initiatives may be launched without clear responsibility for maintenance, monitoring, or adaptation as environmental conditions evolve.

These challenges highlight the importance of institutional learning and adaptability. Resilience-oriented planning requires governance systems capable of revisiting assumptions, adjusting strategies, and responding to unforeseen outcomes. The

Saudi experience illustrates that without such capacity, even well-intentioned planning initiatives can generate unintended consequences. Addressing the implementation gap therefore requires not only stronger coordination mechanisms, but also a rethinking of how planning institutions are empowered, resourced, and evaluated.

9.7 What the Saudi Experience Reveals for Arid Urban Futures

While firmly grounded in the Saudi context, the insights generated by this book extend beyond national boundaries. As climate change intensifies, an increasing number of cities worldwide are confronting conditions of heat stress, water scarcity, and environmental uncertainty that resemble those long experienced by desert cities. In this sense, Saudi Arabia offers a valuable analytical lens for rethinking urbanization under constraint.

The book does not present Saudi cities as models to be replicated. Instead, it positions them as sites where the limits of conventional urban planning paradigms are rendered visible. The challenges identified, reactive urbanism, fragmented governance, reliance on resource-intensive systems, are not unique to Saudi Arabia, but they are amplified by aridity. This amplification exposes the weaknesses of planning approaches that assume environmental abundance or technological substitution as default conditions.

By foregrounding environmental constraint, the book contributes to a growing body of scholarship that calls for a reorientation of urban planning theory. It suggests that planning frameworks must be capable of operating under scarcity, uncertainty, and long-term stress, rather than assuming stability or continuous growth. The Saudi experience underscores the importance of context-sensitive planning that recognizes diversity among arid regions while acknowledging shared structural challenges.

In this way, the book speaks to broader debates on urban resilience, sustainability, and adaptation. It invites scholars and practitioners to reconsider how planning knowledge is produced, transferred, and applied across contexts. Rather than exporting solutions, it advocates for analytical approaches that begin with environmental and institutional realities, using them as the foundation for planning intervention.

The transformation underway in Saudi cities is unprecedented in scale and speed. Megaprojects reshape landscapes; green initiatives reconfigure public space; governance reforms recalibrate institutional roles. These processes will shape urban futures for decades. Whether they result in resilient urban systems or embedded vulnerabilities will depend less on technological sophistication or architectural spectacle than on the capacity of planning institutions to align scale with environmental constraint, ambition with realism, and innovation with adaptability. This book closes by reaffirming its core premise: urban resilience in arid environments is neither a by-product of growth nor a feature of iconic development. It is a condition achieved when planning systems internalize environmental limits as structuring parameters rather than external obstacles. In this sense, the future of urbanization in Saudi Arabia, and in other climate-constrained regions, rests not on the magnitude of projects undertaken, but on the maturity of planning frameworks that guide them. Resilient urban futures emerge not from singular solutions, but from integrated, disciplined, and context-aware planning.

9.8 Final Reflections

This book provides a rigorous and critical examination of urbanization and resilience in Saudi Arabia by placing planning at the center of analysis. Across its chapters, it demonstrates that urban outcomes in arid environments are shaped not only by economic ambition or policy intent, but by the cumulative effects of planning decisions made under conditions of environmental constraint and institutional complexity. By integrating perspectives on sustainability, resilience, green infrastructure, and megaproject-led development, the book advances a coherent framework for understanding how urban vulnerability and resilience are produced over time.

The contribution of the book lies in its ability to bring together fragmented strands of scholarship into a unified, planning-led interpretation of Saudi urbanization. Existing literature has often addressed elements of this story in isolation, focusing on individual cities, sectors, or policy initiatives. This book moves beyond such fragmentation by examining how planning, governance, and environmental limits intersect across scales and over time. In doing so, it fills a notable gap in contemporary scholarship and establishes a foundation for future research on arid and climate-constrained urbanization.

The book also positions itself as a reference for both research and teaching. For scholars, it offers a set of analytical lenses and empirical insights that can inform comparative studies and theoretical development within urban planning and resilience research. For educators, its structured progression from conceptual framing to thematic analysis and empirical investigation supports its use across undergraduate and graduate curricula. At a time when integrated teaching materials on urbanization in arid environments remain limited, the book provides a context-specific, theory-informed resource that bridges abstract concepts and real-world planning challenges.

Importantly, the book does not advance prescriptive solutions or universal models. Its contribution lies instead in fostering critical engagement with the assumptions, trade-offs, and institutional realities that shape urban development under constraint. By emphasizing planning as a mediating practice, it highlights the importance of long-term thinking, coordination, and adaptability in navigating uncertainty. These qualities are essential not only for Saudi cities, but for urban regions worldwide as they confront intensifying climate pressures.

This book argues that rethinking urbanization in arid environments requires moving beyond inherited planning paradigms and toward approaches grounded in environmental realism and institutional awareness. The Saudi experience demonstrates that resilience is not achieved through ambition alone, but through planning systems capable of learning, adapting, and operating within limits. As climate change reshapes urban conditions globally, the insights offered here invite a broader re-examination of how cities can be planned and governed in an increasingly constrained world.

Index

A
Arid cities, 7, 16, 139–146, 149–154, 171, 172, 174
Arid environments, 5–8, 11, 13–15, 139, 141, 142, 151, 153, 154, 160, 161, 164, 168, 171, 173, 175, 178–180, 182, 224–230

C
Circular economy, 116, 122, 186–189, 208, 212
Climate adaptation, 9, 13, 15, 25, 122, 124, 137–139, 142–144, 146, 147, 150, 153, 154, 182
Climate change, 3–5, 7, 8, 11, 13, 22, 23, 30, 34, 35, 37, 40, 49, 50, 53, 55, 58, 59, 62, 63, 76, 82, 83, 92–94, 103–107, 109, 111–113, 115, 117, 119–124, 127–129, 172, 185–187, 189, 207–213, 225, 228, 230
Climate constraint, 143
Climate-constrained development, xi
Climate stress, 9, 143, 150, 151, 154, 162, 165, 171, 174, 175, 183, 226

E
Environmental challenges, 34, 49, 53, 86, 91, 96, 179, 187
Environmental constraint, 5–9, 13–15, 138, 140, 147, 148, 150–154, 161, 162, 165, 167, 168, 170, 171, 173, 174, 177, 178, 180–183, 223, 224, 226–229

G
Governance and planning capacity, 162
Green infrastructure, 7, 9, 11–15, 111, 120, 137–142, 144–154, 182, 183, 186, 226, 227, 229

L
Large-scale urban transformation, 160, 170, 181

M
Megaprojects, 9, 12–15, 154, 159–183, 226–229

P
Population growth, 4–6, 8, 19, 28, 30, 41, 49, 50, 62, 65, 82–85, 103, 192, 197, 208, 209, 223

S
Saudi Arabia, 5–16, 20, 21, 23–36, 38–41, 50–53, 55–60, 62–66, 76–86, 88–96, 104, 106, 109–111, 113–116, 118–122, 124–129, 137–139, 141, 143–150, 152, 154, 160–176, 178, 180–183, 186–191, 197, 203–206, 209–211, 213, 223–229
Sustainable development, 3, 6, 20–23, 25–28, 30, 34–36, 38–41, 49, 50, 53, 55, 62–64, 66, 76–78, 82–84, 95, 96, 103, 121, 128, 129, 188, 189, 207, 212

A. I. Almulhim and P. B. Cobbinah, *Urbanization Rethought*, Urban Sustainability,
https://doi.org/10.1007/978-981-92-0118-1

Sustainable urban development, 5, 20, 22, 23, 25, 28, 31, 38, 39, 51, 77, 95, 124, 187, 188, 192
Sustainable urbanization, 3–5, 20, 21, 28, 35, 36, 39–41, 65, 66, 76, 77, 79, 81, 82, 85, 88, 90, 91, 93, 95–97

U
Urban growth, 6, 7, 10, 14, 15, 19, 21, 25, 30, 31, 35, 49, 51, 53, 57, 60–62, 77, 81, 86–89, 95, 103, 104, 109, 137, 154, 182, 191, 213, 225
Urbanization, 3–10, 12–16, 19–28, 30–41, 49, 50, 52, 53, 55–66, 75–78, 80–86, 88, 90–97, 103, 104, 109, 120, 128, 129, 137, 139, 150, 154, 160–162, 166, 169, 171, 177, 180, 182, 183, 185–189, 192, 193, 197, 204, 206, 208–213, 223, 225, 226, 228–230
Urban nexus, 169, 187–190, 192, 205–210, 212, 213
Urban planning, 4, 5, 9, 11, 13–15, 20–23, 25, 28, 31, 32, 38–41, 75–77, 79–86, 88, 90, 91, 93–97, 104, 105, 107, 109, 111, 112, 116, 128, 137–147, 152, 165, 166, 176, 188, 189, 206–213, 224, 226, 228, 229
Urban policy, 10, 11, 19, 34, 38, 39, 104, 107, 110, 113, 127, 129, 159, 185
Urban resilience, 5–7, 11–16, 82, 83, 104–107, 110, 111, 115, 127–129, 137–140, 142, 144, 146, 147, 150–154, 160–164, 166–168, 170, 174, 177, 181, 185–189, 205, 208–213, 224–226, 228
Urban sustainability, 7, 23, 34, 49, 52, 53, 64, 75, 76, 89, 153, 159, 165, 185–188, 207–209

W
Water scarcity, 5–8, 10, 11, 14, 16, 34, 57, 93–95, 137–139, 141–148, 151, 153, 163, 164, 166, 168, 170, 171, 173–175, 181, 182, 186, 223, 225, 226, 228

The manufacturer's authorised representative in the EU is Springer Nature Customer Service Centre GmbH, Europaplatz 3, 69115 Heidelberg, Germany. If you have any concerns regarding our products, please contact ProductSafety@springernature.com

Printed and bound by CPI Group (UK) Ltd, Croydon, CR0 4YY
21/09/2026
02219986-0003